HEALERS

A CIVIL WAR SAGA

STEWART COHEN, PH.D.

Publisher's Information

EBookBakery Books

Author contact: stewcohen@yahoo.com

ISBN: 978-1-938517-43-3

TABLE OF CONTENTS

The Horror of War

"I cannot give you an idea of the terrors of this battle. For ten long hours, it literally rained ball, shells, and other missiles of destruction. The sight of the dead, the cries of the wounded, the thundering noise of battle can never be put on paper. The dead, the dying and the wounded all mixed up together. Friend and foe embraced in death. Some crying for water. Some praying their last prayers. Some trying to whisper to a friend their last farewell message to their loved ones at home. It was heart rendering."

Jesse Reed, Confederate soldier, 8th South, Carolina Infantry

"Our cause was lost from the beginning. Our greatest victories – Chickamauga and Franklin – were our greatest defeats. Our people were divided upon the question of Union and succession. Our generals were "scrambling" for "Who ranked". The private soldier fought and starved for naught. Our hospitals were crowded with sick and wounded, but half provided with food and clothing to sustain life. Our money was depreciated to naught and our cause lost. We left our homes four years previous. Amid the waving's of flags and handkerchiefs and the smiles of the ladies, while the fife and drum were playing Dixie and the Bonnie Blue Flag, we bid farewell to home and friends. The bones of our brave Southern boys lie scattered over our loved South. They fought for their country, and gave their lives freely for that cause… We shed a tear over their flower-strewn graves. We live after them. We love their memory yet."

"The disposal of the dead posed serious problems for the living. The sheer volume of the bodies to be buried, some unidentifiable, was overwhelming. Yet the task had to be performed. And often, expediency ruled the day. As one soldier described the process:

"… long trenches were dug about six feet wide and three to four feet deep. The dead were rolled on blankets and carried to the trench and laid heads and feet alternating so as to save space. Old blankets were thrown over the pile of bodies and the earth was thrown on top."

G. R. Lee, soldier

July 1863, Gettysburg, Pa.

"We were all glad that the storm had passed, and that victory was perched upon our banners.

But oh! the horror and desolation that remained. The general destruction, the suffering, the dead, the homes that nevermore would be cheered, the heart-broken widows, the innocent and helpless orphans! Only those who have seen these things can ever realize what they mean."

Rampant Killing

"May the heart of this fair land be forever inclined unto wisdom, so that we may never fall into the folly of another war, and be compelled to pay the fearful penalty that is sure to follow.

"For a number of days after the battle, amputating, nursing and cooking continued on the premises, after which the wounded were removed to the different corps' hospitals that Dr. Letterman, (chief medical officer of the Army of the Potomac) had set up in anticipation of this great battle. During this time many a brave and noble spirit went from its tenement, and passed to the great beyond. This is what it meant, when they silently carried out a closed rough box, placed it upon a wagon and drove away."

Editor's Note: Excerpts drawn from the impressions of observed events, years after the battle, as penned by Tillie Pierce a young resident of the town of the occurrences that took place in Gettysburg, Pennsylvania in July of the summer of 1863.

PREFACE

A Letter Home

A bloody path of death followed the armies of the North and the South as they marched across the land. Not only did the killing continue as the war progressed, but in some cases, increased as well. On the evening of May 10th, 1864, as the Civil War entered its fourth year, 26-year-old James Robert Montgomery, a private in the Confederate Signal Corps in Virginia, wrote a letter to his father back home in Camden, Mississippi. At the time Pvt. Montgomery was dripping blood on the paper he was using as he wrote his last letter. He was dying. And he knew it. His injury stemmed from a horrific arm wound he had sustained just a few hours earlier that as he realized, would soon kill him. The letter, his last testimony, read as follows:

"Dear Father.

This is my last letter to you. I have been struck by a piece of shell and my right shoulder is horribly mangled and I know death is inevitable. I am very weak but I write to you because I know you would be delighted to read a word from your dying son. I know death is near, that I will die far from home and friends of my early youth but I have friends here too who are kind to me. My friend Fairfax will write you at my request and give you the particulars of my death. My grave will be marked so that you may visit it if you desire to do so. It is optional with you whether you let my remains rest here or in Mississippi. I would like to rest in the graveyard with my dear mother and brothers but it's a matter of minor importance. Give my love to all my friends. My strength fails me. My horse and my equipment will be left for you. Again a long farewell to you. May we meet in heaven."

Your dying son, *J. R. Montgomery*

James Montgomery's friend, Fairfax, did write soon thereafter to his friend's family -- forwarding some of his effects -- and assuring his father that he had been conscious to the end, and that his son had died at peace with himself and his maker. But it was of little consolation to the Montgomery family. Though his grave had been marked at the time, the exigencies of war prevented its discovery. Despite their best efforts at recovery of his remains, the location of body remained hidden; his family was never able to find it, and they were never able to realize their fond hope of bringing their dead son home. It was one of the many unforeseen and unfortunate consequences of a war that, at the time, but not the case as assumed by some, was to be short and minimally destructive. Why some said that it would be over in three months. But, of course, it was not. The war lasted four years, with its legacy and its remnants still with us today.

A Description of the Fallen:

"Lying upon the ground with no shelter from the fierce heat of the sun by day or the dew at night, were some three hundred rebel wounded. They had as yet received no care from surgeons. Many of them were in the most horrible condition that the mind can conceive. Some were shot through the head, body, or limbs, others mangled by fragments of shell, and all suffering the greatest torment. We gave them water, and shared with them the contents of our haversacks, but there was nothing else we could do. Words are powerless to convey an adequate ideas of these harrowing scenes."

Sergeant Wilbur F. Hinman, 65th Ohio Infantry

More Messages

July 14, 1861, Camp Clark, Washington

"My very dear Sarah:

"The indications are very strong that we shall move in a few days—perhaps tomorrow. Lest I should not be able to write again, I feel impelled to write a few lines that may fall under your eye when I shall be no more . . . .I have no misgivings about, or lack of confidence in the cause in which I am engaged, and my courage does not halt or falter. I know how strongly American Civilization now leans on the triumph of the Government and how great a debt we owe to those who went before us through the blood and sufferings of the Revolution. And I am willing—perfectly willing—to lay down all my joys in this life, to help maintain this Government, and to pay that debt . . Sarah my love for you is deathless, it seems to bind me with mighty cables that nothing but Omnipotence could break; and yet my love of Country comes over me like a strong wind and bears me irresistibly on with all these chains to the battle field.

"The memories of the blissful moments I have spent with you come creeping over me, and I feel most gratified to God and to you that I have enjoyed them for so long. And hard it is for me to give them up and burn to ashes the hopes of future years, when, God willing, we might still have lived and loved together, and seen our sons grown up to honorable manhood, around us. I have, I know, but few and small claims upon Divine Providence, but something whispers to me—perhaps it is the wafted prayer of my little Edgar, that I shall return to my loved ones unharmed. If I do not my dear Sarah, never forget how much I love you, and when my last breath escapes me on the battle field, it will whisper your name. Forgive my many faults and the many pains I have caused you. How thoughtless and foolish I have often times been! How gladly would I wash out with my tears every little spot upon your happiness . . .

"But, O Sarah! If the dead can come back to this earth and flit unseen around those they loved, I shall always be near you; in the gladdest days and in the darkest nights . . . always, always, and if there be a soft breeze upon your cheek, it shall be my breath, as the cool air fans your throbbing temple, it shall be my spirit passing by. Sarah do not mourn me dead; think I am gone and wait for thee, for we shall meet again . . .”

Sullivan Ballou was killed a week later at the first Battle of Bull Run, July 21, 1861. He was 32 at the time of his death.

Official Notification

It was a dreaded, but all too common letter, one appropriate to the time and place of its origin. And unfortunately, as its recipients knew, it was a notification that had appeared all too often across the innumerable battlefields in which the struggle was now in its second year. While occasion dictated something of its specific content, it was not uniform in its description of the events surrounding its occurrence, only its diminishing meaning. A loved one, a husband, a father, a son, a bother, all of these people had fallen. That could not be mistaken.

The length of the message varied as well as the details reported. It was a letter sent home, a communication usually penned by a colleague or an officer of a deceased soldier to offer precious information and sparse comfort to a grieving recipient or family member awaiting word of the fate of a loved one. Such letters were similar for both armies; moreover, their common content was like many others of their kind, designed to explain both the circumstances surrounding a soldier's untimely demise to his family, where possible, and his experience of the last rites that were administered, if here too circumstances presented themselves, allowing the writer and his colleagues to do so at the time of his death.

President Lincoln's letter to Mrs. Bixby

In the fall of 1864 Massachusetts Governor John A. Andrew wrote to President Lincoln asking him to express condolences to Mrs. Lydia Bixby, a resident of his state and a widow whom, at the time, was believed to have lost five sons during the Civil War. The Boston Evening Transcript printed Lincoln's letter to her. Later, it was revealed that only two of Mrs. Bixby's five sons died in battle (Charles and Oliver). Of the remainder of her family, one deserted the army, one was honorably discharged, and another deserted or died a prisoner of war (cause unknown). The letter written in an attempt to ease the grief of their mother read as follows:

Executive Mansion, Washington, Nov. 21, 1864

Dear Madam,

"I have been shown in the files of the War Department a statement of the Adjutant General of Massachusetts that you are the mother of five sons who have died gloriously on the field of battle.

"I feel how weak and fruitless must be any word of mine which should attempt to beguile you from the grief of a loss so overwhelming. But I cannot refrain from tendering you the consolation that may be found in the thanks of the Republic they died to save.

"I pray that our Heavenly Father may assuage the anguish of your bereavement, and leave you only the cherished memory of the loved and lost, and the solemn pride that must be yours to have laid so costly a sacrifice upon the altar of freedom.

Yours, very sincerely and respectfully,
A. Lincoln

This missive, unfortunately, was one of many similarly written at the time, aside from those where there would be no letter, no acknowledgment, no statement of condolence, nor any recorded account made otherwise. Many of those soldiers that died and reported as such were "fortunate" in a limited sense, as circumstances allowed, to have their fate known and conveyed to their loved ones or to be acknowledged by their colleagues in their final hours. For some, if their families could afford the cost, relatives of the deceased would come and search for their remains in order to bring them home. Some of these searches lasted many years beyond the war. But more often than not, this was not the case for most families. Their resources were limited or lost during the war. Nor were such discoveries easily made or were they likely to be.

Many soldiers sharing their fate, but less fortunate as events transpired were neither granted similar account, comparable recognition, nor postmortem honor. These men, most unclaimed by their colleagues, were often left dying on the battlefield, and were far too great in number and circumstance over the course of this long war to allow an occurrence of honor. Often these poor victims of a ghastly horror resided in body and spirit on open fields and plains. Here they were often subject to the cruel ravages of the elements and jaws of preying animals, in places of unmarked graves, or sites that were largely unknown to them or to their comrades, far from home or familiar places. And in these places over the course of numerous battles their deaths were simply too abundant to count, or they were left to be claimed by the land as armies moved across battlegrounds as the business of war compelled them. And there they remained anonymous and their fate unknown then and even now. Still others, the more fortunate, were remembered in their way.

Ward 10
Lincoln Hospital Washington D.C.
Jan. 14th 1863

"Mrs. Ann Scanlan,

"Dear Madam- its become my very painful duty to inform you that your husband has just breathed his last. He died at 25 minutes to seven o'clock this evening without pain. As quietly as the infant sinks to rest from the bosom of its mother so peacefully did he breathe out his last sigh and resign his spirit into the hands of the God who gave it. I know how great will be your grief upon reception of this sad news but it is the will of God that it should be so and you must try and bear the bereavement with resignation, knowing that is not for you to question His right to do with His own as He sees fit: "The Lord gave and the Lord taketh away: blessed be the name of the Lord!" Every heart knoweth its own sorrow; and there is a grief which cannot be expressed in word, God alone has power to comfort you and to bind up your wounded and bleeding heart, in this your season of great distress; turn then to Heaven and may He who is the refuge of the weary- the hope of the sorrowing of earth, be with you and sustain you in this hour of trial and throw the arms of His everlasting salvation around you! Mr. Scanlan received the last rights of his religion this morning- the Sister, who has charge of this ward, has been constant in her attendance upon your husband and has done all in her power to alleviate his sufferings. I am not myself a Catholic and do not therefore understand the peculiarities of that faith, but Sister told me that the Priest had administered all the last rites necessary in such cases provided. The Priest has seen him several times and was present last Sunday morning and the Sacrament was I believe administered. It will be a great satisfaction to you to know that it is, as it is in this respect I have thought of the difference between the case of your husband and of another poor fellow who died recently and belonging to a different faith. He passed away without so much as having a Minister of Religion near him to breathe a prayer for the peace of his departing soul – such is the difference between the two religions. It has set me to thinking and I shall do so seriously I assure you after this. Your last letter was received, and I offered to write an answer knowing that you would naturally feel anxious to hear from him, but he said he'd wait a day or two first to see if there would be any change for the better. He felt sensible, I think, that his end was approaching for he requested me to make a note of his feelings at that time- this was yesterday forenoon, I think. He did not talk a great deal as it hurt him to do so much. "After I am dead, write to my wife and tell her that I died a natural death in bed, having received the full benefits of my church." "Say that I felt resigned to the will of God and that I am sorry I could not see her and the children once more. That I would have felt better in such a case before I died. It is the will of God that it should not be so, and I must be content to do without." This was about the substance of what he said. I read it to him and he said it was all that would be necessary to write. His pay amounts to some 6 1/2 months not having received any since the 1st of July. This of course you are entitled to draw and you can do so by getting some friend to assist you, understands about it. The few things in this letter are all his personal effects. The rest of his things letters & c. he said to burn-which will be done. I will close for the present.

I remain very truly your well wishes, William Duffie. If you wish to answer this, please direct to me Lincoln Hospital, Washington, D.C., Ward 10. (3)

Another Casualty

At Gettysburg, despite the enormous cost in human life in both armies, the number of civilian casualties was less pronounced. But there was one who was, nevertheless, well remembered.

Jennie Wade, A Tragic Story

Jennie Wade was a young 20-year-old resident of Gettysburg. Before the Battle she had become engaged to Corp. Johnston H. Skelly of the 87th Pennsylvania. In Gettysburg she worked as a seamstress with her mother in their home on Breckenridge Street. During the days surrounding the battle she baked bread and carried water to the soldiers.

For safety during the first day's battle, Jennie and her family moved to the home of Jennie's sister, Georgia Wade McClellan on Baltimore Street. Her sister had just given birth with great difficulty around 2:15 P.M., one hour before the Confederates rode into Gettysburg. At the time Jennie was caring for her sister. The McClellan side of the house on Baltimore Street, less than 50 yards north of Cemetery Hill, thus housed Mrs. Wade, Jennie, her brother Harry, her young boarder Isaac, her sister Georgia, and her newborn son. At that particular moment, there was little heavy fighting in the area. However, a Federal picket line ran behind the little brick house, and there was intermittent skirmishing between the Federals and Confederates in the town. Protected by the sturdy brick walls of the house, Jennie and the others lived for three days in the midst of the battle.

Jennie spent most of July 1 distributing bread to Union soldiers and filling their canteens with water. By late afternoon on July 2, the diminishing supply of bread made it apparent that more bread would be needed the next day. Jennie and her mother proceeded to heed that call. On the morning of the 3rd at about 7 A.M. on the morning of July 3, Confederate sharpshooters began firing at the north windows of the house. The prep work to bake biscuits was begun at 8 A.M. At about 8:30 A.M. while Jennie stood in the kitchen kneading dough, a Confederate musket ball smashed through a door on the north side of the house, pierced another that led into the kitchen, and struck Jennie in the back beneath her left shoulder blade, embedding itself in her corset and killing her instantly. The cries of her sister and mother attracted Federal soldiers who carried her body to the cellar. Later she was buried in Evergreen Cemetery. In the early afternoon of July 4, Jennie's mother baked 15 loaves of bread from the dough that Jennie had kneaded. After it baked she distributed it to the soldiers.

The tragedy of Jennie's death unfortunately, had been extended to others in her circle, her fiancé Corp. Skelly had been wounded and taken prisoner at Winchester on May 13. From there he was transferred to Virginia. However, after a month, he died in a military hospital on July 12. His family learned of his death only several days after the Army of Northern Virginia had withdrawn from Gettysburg to turn South.

1

THE BEGINNING

M odern medicine, as we know it, began in war. As war came it ush-
ered in a vast array of modern weapons, faster modes of transportation, and new ways of communicating over vast distances; many facets
of life at the time and even practices that have endured over time were
profoundly influenced and most often changed. These changes extended
to medical practice, both of necessity and to the call of common human-
itarianism.

The American Civil War was the first modern war. Innovations in
telegraphic communications, the growth of railroads that now allowed
the rapid transport of troops via systems of interlocking connections, the
production of more accurate, more deadly, and more and greater rapid
firing weapons, all combined in their sudden and horrific amalgam to
bring death, injury and massive destruction to the towns and cities of the
nation. But battlefield injury alone did not contribute to the vast carnage,
or as heavily to increases in the morality rate, as would be anticipated.
Rather, it was the widespread occurrence of disease and the failure to
adequately attend to wounded soldiers that did so. And in so doing this
war contributed to, as well as maintained a horrendous tradition, one that
has prevailed in wars fought since the beginning of time.

When the Civil War began the military and civilian medical personal
needed to attend to its awful consequences. However, like other facets

of this war they were ill - prepared and poorly trained to address the myriad of issues and demands that they would face. Nor were they able to anticipate or meet the challenges, or if truth be told, of the myriad and unimagined horrors that would follow: the war would change the lives of the soldiers that fought and those of millions of their fellow citizens, for generations, as well.

Some Personal Observations From the Front

The performance of surgery on the battlefield was rough and less than optimal. And it was certainly not for the squeamish nor faint of heart. "I witnessed some terrible things on the battlefield and beyond during the war. And I was there for most of it. But let me start from the beginning.

"I was only twenty-years old when I enlisted with the Pennsylvania Infantry. I was hoping to fight, as many of my fellows. However, when it was discovered that I had completed a year as a medical student, I was quickly shifted to the Medical Corps. That is, what at that time there was of it. No one knew how long this war would last or how many lives it would take before it was over. So, there was great anticipation, but little preparation."

For the physicians that would serve the war offered neither glory nor honor. It was a savage and brutal conflict inflicting punishment never before seen that they were forced to repair, where and when they found that they could.

"A surgeon, this is what the military called field doctors gained their prominence by using the most dramatic methods of treatment and acting with desperate speed in addressing the needs of the wounded. The most often employed practice in surgery was amputation, a most ghastly and bloody procedure.

"The work of the surgeon is gory, often brutal. Large numbers of wounded are first treated on the field of battle or at aid stations, these being set up on the edge of the battle field. Men are then transported to field hospitals for surgery. Getting the wounded soldier to the surgeon is often times slow. Surgeons many times need to go out on to the battle ground during the battle and either work on the wounded right on the ground or bring them back to the field hospital and work on them there.

The Civil War has seen the beginning of the ambulance on the battlefield. It also bore witness to the initiation of triage, where the selection of who among the wounded would get treated first. In the aid station or field hospital only the wounded that required the surgeon would see him. Minor injuries are treated by either the assistant surgeons or by stewards.

"Let me tell you, the tools of these surgeons varied from unit to unit, some better equipped than others. Most often the surgeons were required to bring with them their own surgical kits." Most of these kits had the following basic tools as depicted in the rough photographs shown below:

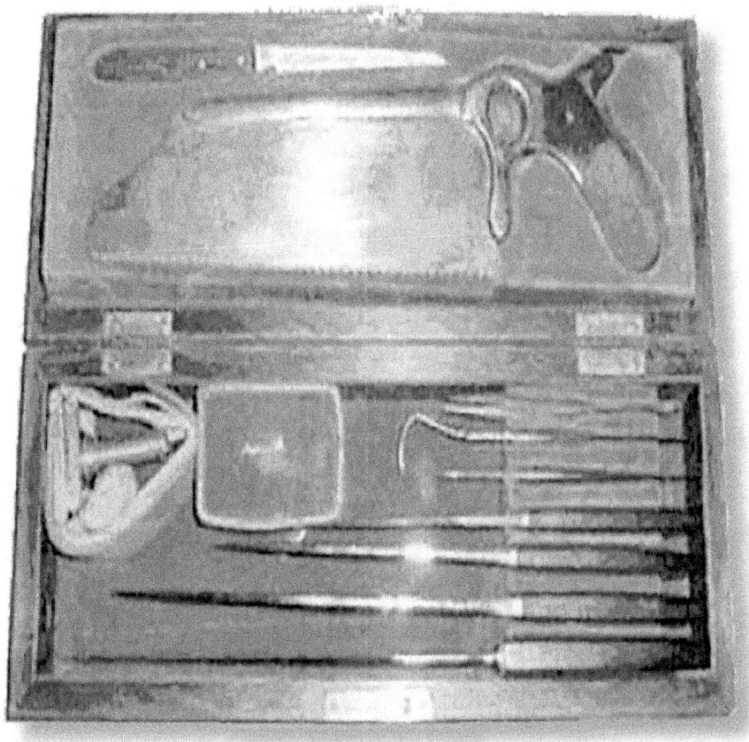

"CAPITAL SET" FROM A SURGEON'S KIT

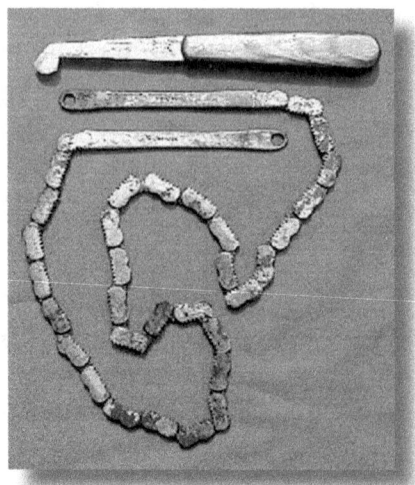

"CHAIN SAW" WITH A FOLDING SCALPEL.

Scalpels, scalpel guides, amputation knives (used to prepare soft tissue prior to amputation) various amputation saws, tourniquets, the ever-necessary bullet probes for locating wounds and extractors for removing bullets and shrapnel were part of the medical kit. These were the most often used "tools".

"The surgery that I assisted with was most often carried out in the open air, usually in full view of anyone passing nearby. We worked quickly. Body parts taken off by amputation, were tossed aside, and often amass in huge piles. By that time, flies, insects of all kinds were drawn to the discarded body parts, and the blood of the wounded and dying."

"Once on the surgical table when amputation was warranted a wounded soldier would be restrained by stewards or others, as was necessary. At times anesthetics were available in the form of chloroform, or ether to ease the burden of surgery. But supplies were limited, and needed to be employed sparingly. Morphine, opium pills, and Laudanum were used for the control of pain where they were available. However, in most case these anesthetics were in short supply throughout the war and were not always available.

"Chloroform or ether was dripped from a bottle or a can onto a cloth mask held close to the patient's face. Ether, being a volatile liquid could burn the patient if it came into contact with his skin. Moreover, the fumes could cause the medical staff to become sick or sleepy as they worked unless the area was well ventilated, a condition favoring the use of open-air surgeries."

Initially, morphine was used orally, then being rubbed into wounds and later was injected directly. Opium pills the best available pain control medicine was administered in the same form as morphine. Laudanum, a form of opium was also taken orally. All of these pain medicines were addictive, causing other problems as the patient recovered. However, the effects of these medications were unknown at the time.

Typically, an injury, such as a leg wound, would have any clothing removed above and below the site of wound. If the bone was not broken then the wound could be probed to see if the bullet or shrapnel was still in the body. If an object was easily located and could be reached with the extractors this procedure was performed and the wound would be sewn closed. An assistant surgeon would most commonly perform this simple procedure. For more difficult removal of foreign objects the surgeon would cut a path into the wound preparatory to extraction. Once the object was removed the surgeon would move to another patient and an assistant surgeon would repair the wound. Any bleeding during the surgery would be soaked up with rags and sponges. Lacking knowledge of the causes of infection, these same rags and sponges, unfortunately, would then be used on the next patient along with the same scalpels, probes, and extractors. The danger of infection, an all too frequent occurrence, and its varied consequences, usually amputation and death, resulting from these practices was, as might be anticipated, extraordinarily high.

As one observer noted, "Because of the way surgeons worked it is no wonder that a number of uncomplimentary nicknames such as "saw bones" came about. It was a very dirty, very gruesome business that the dedicated men and women and I, myself encountered to aid the soldiers. At first I found the work deplorable. Over time however, I found grace in my work. And I took inspiration from these men, and was only then that I was able to assist in healing them."

2

LETTERMAN ENTERS THE WAR

The United States Medical Military Corp, at the initiation of the Civil War in 1861 consisted of less than 100 surgeons and fewer assistant surgeons. Quickly, this small pool of insufficient numbers of men, moreover, was soon depleted, especially so as men of different persuasion took up different sides in this conflict. Medical personal were not immune to the political arguments of the time nor their affiliations with one side over the other; rather, men with dramatically different points of view or possessing different regional allegiances resigned their medical posts in the regular army and chose instead to follow the destiny of their chosen preference.

In support of the medical needs concurrent with the war, hundreds of civilian doctors on both sides enlisted to help fill the void created by this paucity of medical personnel. Consequently, these "grafted" doctors, as they were called, untrained in military matters and lacking experience in the care and treatment of men subject to war-affiliated damage that were placed under their care, had to be taught the ways of the military and the unique demands that was posed in attending to their charges.

The Civil War was fought before the idea of medical practice incorporating wartime demands of such trauma was an integral part of the physicians training or curriculum. While there were medical schools at the time, they were inadequate to the task. Nor did they sufficiently or adequately address trauma medicine of the magnitude and scope of the problems create by the severity of injury caused by the Civil War. Many training centers for doctors required students to attend one or two years of lectures. Moreover, the program of studies of a student enrolled for the full curriculum, usually a 2-years program, was limited; enrollees would often sit in on the very same lectures during their second year of study as they had attended during their first year.

There was at the time an alternative route to becoming a physician. This path entailed, as many professions at the time, an apprenticeship consisting of on-the-job training under the guidance of a master practitioner. A person wanting to become a doctor would work with another medical practitioner, until his mentor felt that his student had learned all that could be taught. At such time the apprentice would be sent out to start his own practice. Needless to say, the practice of the medical arts was limited to men. In spite of the need, there were no schools of nursing, nor the training of women in the caring arts at the onset of the War, at least until some time later, when the need for more attendant care, caused by the large unanticipated number of casualties demanding more care, became evident.

By1860 there were over 100 medical schools and the majority of doctors trained had some type of formal education. Yet, neither learning through direct experience nor formal education could prepare the doctors of the 1840's and 1850's for what was to besiege them the next decade in the Civil War.

The War offered a form of "on the job training" for everyone who worked in the medical field. Clearly, the challenge was substantial and most often, overwhelming; there were few men, or women, for that matter, that had sufficient experience to handle a war like this one. Aside from those few doctors that had served in the Mexican War in the previous decade, experience in the field was lacking; moreover, there was no training or protocol to follow nor venue to assist in that vital task. Neither was there a bank of experiences nor medical traditions available from previous engagements of the sort present among those people that would serve.

The numbers and forms of casualties that occasioned the circumstances of the war exceeded our ability to repair ourselves. Of nearly 3 million soldiers who fought in the war, over 750,000 known or accounted for would die during this time, others would suffer severe and debilitating wounds, or would be unaccounted for thereafter. Around 400,000 died from disease, while 218, 000 died in battle or from wounds received during battle. Among the dead, it is believed that the North lost 359, 528 men and the South had lost 258,000 men. These figures, of course, are approximate since more accurate records were neither systematically accumulated nor properly catalogued. Furthermore, in the heat of battle such accounting was difficult and/or haphazard. Many wounded men that were not directly killed in battle and often seeking aid might escape the fighting or be taken from the battlefield. However, unattended they would often succumb shortly thereafter and be unaccounted for after the fighting in a given sector terminated, or moved on to a more distant place.

Acquiring information about the fate of a loved one by his family was often a limited venture; identities were often unavailable as well as not readily attainable. Many dying soldiers expressed the thought that they would hope to be buried at home. Yet, as armies moved over vast distances thousands of fallen soldiers were left in places where they had fallen or were unaccounted for, whether by the place where their bodies often lay unattended or uncared for; unfortunately, knowledge of the identities of these men would not be determined then or beyond. The battlefields were too widely scattered across the country and the men that marched to battle across that "foreign" terrain could only hope to be remembered in memory.

For the lucky few that were rescued, or removed from the fields of conflict adequate care was often neither in place nor assured; most "treatment centers" were informal or haphazard, dependent on the resources of the communities where the fighting had occurred. Moreover, most often in these places the resources for caring for the injured were limited and inadequate. At Gettysburg a town of 2,300 persons at the time of the battle, casualties (killed, missing, or captured) reached over 52,000 at best count.

In many places, the hospitals, where they existed, were often in disarray and poorly organized. The many deficiencies that prevailed included insufficient and poor housing, inadequate sanitation and haphazard nutrition. Moreover, personnel that staffed these facilities were untrained in the latest medical practices and techniques and in some cases, offered care in the hope of achieving profit or personal benefit. Finally, an existing system permitting medical accountability and proper record keeping necessary to log the tragedies of war was absent; there was no systematic nor effective accounting of the human costs of the war during the various phases of the conflict.

Infection and Disease.

Compounding the problems produced by injuries caused by the war, an understanding of antisepsis in the treatment of the infirmed was just becoming known in America. Moreover, that sphere of medicine was influencing the frontiers of medical practice in other places, primary Europe, where the Crimean War (1853-1856) was recently fought, not the United States. This lack of knowledge concerning the role of infection and its prevention in the causes of illness resulted from the infrequent and haphazard employment of basic hygienic practices and sanitary principles. Among common practices, participating doctors in a rush to treat the men placed under their care would seldom wash their instruments or their hands (or replace and discard used bandages among the men), extending opportunities for infection and disease to occur, while attending to different patients. Moreover, after surgical repair had been performed, their patients were generally housed in ill-prepared and over crowded quarters, while food and water was often contaminated by fecal material produced by the men and their animals. With overcrowded conditions inherent in such arrangements, the spread of infectious diseases was rampant and devastating.

The consequences of poor practice were neither exclusive nor confined to one army. The effects of disease were overwhelming in both armies, North and South, varying in effect, with some units significantly decimated on both sides. It was said that measles cut through the ranks of the 15th Alabama (of later fame at Little Round Top on July 2, 1863 at Gettysburg) at their encampment like a biblical plague or the medieval Black Death. No one, including the small number of surgeons assigned to the army, knew that the disease was air-borne, carried on droplets through the air and that proximity to the virus meant almost certain infection. In this respect, it is somewhat miraculous that the entire Confederate camp at Pageland was not stricken with the disease. Infected soldiers experienced high fever, rash, runny noses, watery eyes, and coughing. Lacking a vaccine and effective treatments, few men who were infected survived the illness. After the initial symptoms appeared, their condition generally worsened. Some soldiers came down with pneumonia and encephalitis (brain inflammation) as a result of measles; others suffered middle-ear infections, severe diarrhea, and convulsions. The worst cases, and there were hundreds of them among the troops of the 15th Alabama, resulted in death.

The first man in the regiment to die was Andrew J. Folmar, 18, a private in Company I. Then many others quickly became sick and had no strength or immunity to fight off the overwhelming disease. About 100 of the regiment's men died over the span of six weeks. A military funeral and burial were performed for each death, and accompanying testimony soon became part of the camp's daily routine. Overcome with emotion from this profusion of sickness and death, one private wrote in despair: "Beneath the soil of Prince William [County], now slumber in quiet repose, secure from summer's heat and winter's cold, from the cares of life and shock of strife, the noblest and best of the regiment."

Those who fell to sickness were stricken by the fear—and the near certainty—of approaching death. Sick and well alike yearned for the comforts of home and to be magically transported from this strange land where so many men were dying. For those on death's doorstep, the longing for home was even more pronounced. "The thought of home is ever uppermost in the mind," admitted one Alabaman, "and a wish exists to be buried with their fathers and the companies of their youth." Their wish would not be granted. At Pageland, the "Dead March" was so frequently heard that men became inured to it and soon did not even inquire as to who had died or was being buried. The endless deaths produced a "crude shock" among the men of the 15th Alabama and, as anyone might expect, "threw a gloom" over the camp that could not be shaken off. So many men were sick that the routine camp duty for those who remained healthy became more strenuous than ever, for now there were fewer hands to do the work.

Throughout the desolation of this epidemic, the 15th Alabama—just like all the other regiments—was ordered to keep up its drill four hours a day, although those who were not sick began to lose their strength under the physical burdens they had to bear.

Colonel William Oates became outraged at the desperate situation transpiring before him. He faulted the army for keeping the sick in the same camp with the healthy men, which ensured that those who were not yet sick soon would be. Years later he wrote in anger:

"I do not know who was responsible for it, but it was a great mistake. There was not that care taken of the men of any regiment, so far as my observation extended, which foresight, prudence and economy of war material—leaving humanity out of the question—imperatively demanded...Had the Confederate authorities made more persistent efforts than they did, hospitals could have been more established in sufficient numbers to have saved the lives of hundreds and thousands of good men, which were for the want of them unnecessarily sacrificed."

Innovations in battlefield medical care came about in the early part of the Civil War in response to the horrendous conditions wounded soldiers faced. A classic example was the Battle of Second Manassas (Second Bull Run), a clash in late August 1862 in which Confederate troops defeated Federal forces outside of Washington, D.C. In that engagement more than 22,000 soldiers were killed or wounded in the conflict—nearly 14,000 of them Federal troops. In the aftermath of the battle wounded soldiers were strewn about the battlefield, crying out for water and medical attention. But there was little comfort for those needing care. Left on the open field to fend for themselves, and without necessary attendant care, their shrieks and moans increased as the days wore on. A full week would pass before all of the Union injured were removed from the battlefield and transported to hospitals—grisly proof of the Army's appalling inefficiency in dealing with casualties.

Often, even when a soldier was rescued from the battlefield, his chances of survival depended upon the effectiveness of the treatment that was provided in the hospital where he was taken. Caretaking often proved to be ineffective and, occasionally, worse. As nurse Mary Livermore reported, "I can conceive that it may be easy to face death on the battlefield…But to lie suffering in a hospital bed for months, cared for as a matter of routine and form, one's named dropped, and only known as "Number 10", "Number 20", or "Number 50"; with no companionship, no affection, none of the tender assiduities of home nursing, hearing from home irregularly and at rare intervals, utterly alone in the midst of hundreds; sick, in pain, sore-hearted and depressed, I declare this requires more courage to endure, than to face the most tragic death."

Something had to be done. However, there were few men with the skill and experience to take up the task. Yet, fortunately, there were a few equal to the undertaking. One of these men was Dr. Jonathan Letterman, a physician from Pennsylvania. Over time, he and several others were able to face and meet the challenges posed by wartime medicine.

The existing structure of the Army's Medical Corp at the start of the War was abysmal. However, that was about to change.

Miraculously, the outlook for the wounded changed within two weeks after Jonathan Letterman's arrival at the scene of the latest fighting. On September 17, 1862, Federal and Confederate forces clashed once more, this time in Sharpsburg, Maryland, at the Battle of Antietam. After 12 hours of fighting, some 23,000 men (over 12,000 Union soldiers and more than 10,000 Confederates) had fallen—the bloodiest single day of combat in the War, and in American history. In stark contrast to the delays and inadequacies experienced at Manassas earlier, however, every injured Union soldier was evacuated from the Antietam battlefield within 24 hours. This change can be attributed to the appointment of Dr. Letterman, the new chief of medical services, physician from Pennsylvania. Dr. Letterman's appointment to this post did not come too soon.

Sir:

You are detailed for duty with the Army of the Potomac as Medical Director. In making this assignment, I have been governed by what I conceive to be the best interests of the service. Your energy, determination, and faithful discharge of duty in all the different situations in which you have been placed during your service of thirteen years, determined me to place you in the most arduous, responsible and trying position you have yet occupied.

W. A. Hammond, Surgeon-General

As the Surgeon-General reported, the medical situation was perilous. Moreover, he was not alone in his concern. In his report of prevailing conditions, General George B. McClellan also alluded to the depth of the problem. He noted that, "the nature of the military operations had also unavoidably placed the Medical Department in a very unsatisfactory condition. Supplies had been almost exhausted or necessarily abandoned; hospital tents abandoned or destroyed and the medical officers deficient in numbers or broken down by fatigue."

At the time of Letterman's appointment the Army of the Potomac had suffered heavy casualties at Harrison's Landing on the James River, where the army had retired after McClelland's failed service in the Peninsula Campaign of 1862.

Major Jonathan Letterman

The recipient of the this critical appointment was a lanky, bearded, man, Major Jonathan Letterman, a 37-year-old military surgeon who'd been named medical director of the Army of the Potomac three months earlier. No stranger to the challenges of organization, Letterman's genius revealed itself the moment he took up his new duties. He immediately instituted revolutionary improvements in battlefield casualty management, and at Antietam, his reforms began to pay off. Letterman's innovations were so farsighted that many are still used today, earning him the title of the Father of Battlefield Medicine.

3

LETTERMAN'S MEDICAL APPLICATIONS

When Letterman was appointed to the Army of the Potomac, it had suffered heavy casualties at Harrison's Landing on the James River where the army had retired after the Peninsula campaign. Dr. Letterman's previous service on the frontier and Indian expeditions into the Western Territories had exposed him to the trials and tribulations of military life. It also gave him insight into the personal needs and requirements of the soldiers.

Letterman intended to make the resources the soldiers so desperately needed available on a larger scale than ever before. His concern was underscored by those expressed by General McClellan. In a report on the subject McClellan wrote, "the nature of the military operations had also unavoidably placed the Medical Department in a very unsatisfactory condition. Supplies had been almost exhausted or necessarily abandoned; hospital tents abandoned or destroyed and the medical officers deficient in numbers or broken down by fatigue."

Added to the difficulties, the situation was especially dire since the Union Army entered the war with only 98 medical officers. This situation was exacerbated further since more than half of the medical professionals affiliated with the army resigned to join the South after the start of the war. These doctors left due to their conflicting loyalties, and in part, because, at the time, promotion was based solely on seniority and not merit or related considerations; these practices added to the frustrations that many of those doctors in service experienced and persuaded them to leave.

Despite these and other disadvantages, Dr. Letterman focused most of his attention on the removal of the sick and wounded from the Peninsula, enforcing sanitary measures for improving and maintaining the health of the soldiers and on providing medical supplies. His orders were concise and practical and even his superior officer General McClellan praised Letterman's efforts in his report: "All the remarkable energy and ability of Surgeon Letterman were required to restore the efficiency of his department; but before we left Harrison's Landing he had succeeded in fitting it out thoroughly with the supplies it required, and the health of the Army was vastly improved by the sanitary measures which were enforced at his suggestion."

4

RESTRUCTURING AND RETRAINING

The Call for Improvements in Medical Practice

"My friend, Lieut. M. is extremely weak and nervous, and the wild ravings of J. C. (another casualty) disturb him exceedingly. I requested Surg. P to have him removed to a more quiet ward." While a reasonable request, circumstances prevented an adequate response to this appeal. Contrary to his plea, the correspondent of this request received the following reply based on the exigencies of the time offered at the time by Emma Edmonds, attendant nurse: "There are five hundred patients here who require constant attention, and half enough nurses to take care of them …While I write there are three being carried past the window to the dead room."

The deficiencies in medical practice at the onset of the War were many. As Surgeon General William Hammond, who was himself recently appointed to his post wrote of one hospital it "defies description". Accordingly, conditions there were "simply disgusting."

He continued, "The outhouses are filled with dirty clothes, such as sheets, bed sacks, shirts, etc., which have been soiled by discharges from sick men. The privy is fifty yards from the house, and is filthy and offensive, ad nauseam. It consists of a shed build over two trenches. No seats; simply a pole, passing along each trench for men to sit on …I do not hesitate to say that such condition of affairs does not exist in any other hospital in the civilized word."

In order to speed the evacuation of wounded soldiers, Letterman created the Union Army's first ambulance corps. Moreover, he set up a uniform system of first-aid stations and field hospitals that brought order to the formerly chaotic, inconsistent treatment of the wounded. He also made sure that medical units received the necessary equipment and supplies. And long before the triage system of medical evaluation was employed in World War I, Letterman instituted standards for prioritizing treatment based on the severity of injuries and the likelihood of survival. Thanks to his efforts, thousands of soldiers survived who might otherwise have died from their wounds.

Born in 1824, Letterman was the son of a prominent Pennsylvania surgeon. He grew up in the town of Canonsburg and graduated from the local Jefferson College in 1845. After graduating from Philadelphia's Jefferson Medical College four years later, he immediately applied for a commission as an Army surgeon. Stationed in California when the Civil War erupted, he returned east at the end of 1861. The following May, he became the medical director of the Department of West Virginia, followed quickly by his appointment as medical director of the Army of the Potomac.

Letterman's improvements also extended to the health of the men. At the time, the Army of the Potomac was in a sorry state, with thousands of sick or wounded troops. Letterman recognized that many of the soldiers were suffering from scurvy, which he quickly cured with increased rations of fresh vegetables. Other problems were more intractable and significantly more difficult to address.

Letterman and his fellow Civil War doctors had to cope with a scale of battlefield casualties that no one had or could have anticipated; the war provided challenges that by breath and scope simply had not presented itself before. The Battle of Second Manassas alone produced twice as many casualties as the entire Revolutionary War. The shocking carnage resulted from a variety of factors including a combination of improved, more deadly weaponry, combined with outdated military tactics in the face of such advancements. Generals on both sides still employed the mass charges of the Napoleonic era, yet in the face of newer weapons, that sort of head-on assault was suicidal. Consequently, it was common for units to lose a third or more of their men in a frontal assault. Moreover, such tactics, employed at Fredericksburg, and elsewhere persisted.

Over 90 percent of Civil War battle injuries came from gunshots, most of those from the ugly half-inch chunk of metal associated with the Springfield musket using the minié ball, a soft lead bullet that could shatter bones and punch a fist-size hole in a man. Such injuries often resulted in the amputation of a limb that could not be repaired in any other manner.

Two other brutally effective projectiles were canister shot and grape shot—cannon shells filled with dozens of small iron balls. When the shells were fired, their casings broke apart, spraying the iron balls into oncoming soldiers in a wall of death, like a giant sawed-off shotgun. The extreme numbers and unusual severity of battlefield casualties taxed the woefully understaffed Army medical corps all through the war. In 1860, the U.S. Army had just over a hundred doctors to treat its 16,000 soldiers. Although the medical corps grew to more than 10,000 surgeons by the end of the war, the ratio of doctors to soldiers actually went down, since the Union Army had swollen to well over two million men by 1865. (Confederate doctors faced an even more daunting situation, with just 4,000 doctors to treat more than a million soldiers.)

Besides having too few doctors, the Army had to rely on physicians with rudimentary skills, often inadequate to the task. Basic medical procedures, previous to the Civil War, hadn't changed much in generations. Doctors usually received only cursory training, often at unregulated two-year medical schools. Moreover, most new doctors had never dealt with gunshot wounds or performed surgery of any sort, including under trying battlefield conditions. Their real training came on the job, often through hacking on the unfortunate soldiers who had ended up on their operating tables. Thankfully, chloroform had come into wide use a few years earlier, and morphine and opium were available to ease soldiers' pain.

However, most Civil War doctors weren't qualified to operate on serious wounds to the head, chest, or abdomen; those injuries were considered fatal and usually left untreated, other than to give the unfortunate victim painkillers when they were available. The most frequent wounds by far were injuries to the extremities, making the amputation of limbs the most common surgical procedure. A practiced "sawbones" could remove a mangled arm or leg in less than ten minutes, using a variety of fearsome knives and saws, while the average amputation often took less time for this procedure where bullet-caused shattered limbs and those subject to gangrene were encountered.

5

CARING FOR THE WOUNDED

In the heat of battle, the typical field hospital looked like a scene from a massacre with surgeons spattered in gore and body parts tossed onto grotesque piles. Surgeons toiled amid the screams of the wounded, the explosions of cannons and the ceaseless rattling of musket fire close at hand. Their bare arms glistened with sweat and blood as they bent over the operating table for hours on.

Moreover, the problems they faced were multifaceted many unknown at the time.

Unaware of the relationship between bacteria and infection, nor of ways to prevent infection, Civil War doctors used the same bandages and surgical instruments over and over without benefit of sterilization, leading to postoperative infections that included blood poisoning, tetanus, and gangrene. Despite the lack of surgical hygiene, however, nearly three out of four amputees survived. Still there were many other ways to die.

Of the 360,000 Union soldiers (at one count) who died in the war, around 220,000 of them succumbed to illnesses such as chronic diarrhea, dysentery, typhoid fever, pneumonia, tuberculosis, smallpox, malaria, and measles. For these varied conditions, the causes of and treatment of infected men were still largely unknown, and therefore untreatable.

AMBULANCE WAGON TRANSPORTING INJURED SOLDIERS

This deadly mix of injury and illness was exacerbated by the haphazard organization of the Army medical corps in the early part of the war. When Jonathan Letterman assumed command as medical director of the Army of the Potomac, medical personnel had no ambulances or dedicated stretcher-bearers. Even worse, there was no coordinated hierarchy for treating the wounded, and doctors ran out of critical supplies far too often. Letterman knew that the key to saving injured soldiers was getting them immediate help. His plan for the Army of the Potomac called for a fleet of wagons that would only be used for transporting the injured, and it required each regiment to assign stretcher-bearers who would have no other combat duties.

Letterman organized the treatment of casualties into three stages: battlefield first-aid stations for stopgap measures, nearby mobile field hospitals for surgery and other emergency procedures, and more distant general hospitals for follow-up care. That simple three-tiered system remains the blueprint for modern battlefield medicine. Finally, Letterman organized a dependable system to provide surgeons with the medical equipment and supplies they needed.

Letterman's comprehensive revamping of the Army of the Potomac's medical corps continued to prove its value after the Battle of Antietam. The new system's biggest test came at the Battle of Gettysburg, which took place July 1-3, 1863, in southern Pennsylvania. The most famous battle of the war involved over 158,000 soldiers—some 83,000 Northern troops led by Gen. George G. Meade and 75,000 Rebels commanded by Gen. Robert E. Lee. The Union victory produced a staggering 51,000 + casualties (23,000 Federals and 28,000 Confederates). Letterman's ambulance corps fielded 1,000 horse-drawn wagons manned by 3,000 drivers and stretcher-bearers. Remarkably, the medical workers were able to clear the wounded from the battlefield by the end of battle on July 4th.

The army was also in desperate need of an ambulance corps. At the time, the "system" for the management of ambulances was neither thorough nor well organized because there was no authority to oversee its use. The ambulances consisted of crude wagons that were often overloaded with wounded men who were tossed about like rag dolls whenever the wagons jolted over the muddy corduroy roads towards the Potomac. These ambulances had civilian drivers who were justifiably terrified of the terrible wounds inflicted on the battlefield and often left their post without picking up the wounded. Observing this occurrence, Letterman immediately drew up a plan for the organization of an Ambulance Corps to evacuate casualties. The soundness of the plan was evident and won the approval by General George B. McClellan. It was adopted, and published with General Orders on August 2, 1862, approximately one month before the battle of Antietam on September 17, 1862.

The plan followed an orderly rationale. First, Letterman formed separate ambulance trains and assigned them to division level commands. Each train consisted of 40 ambulances and was under the command of a Lieutenant. The Ambulance Corps itself was divided into three divisions according to the detachment of troops of each army corp. The use of these ambulances, moreover, was restricted to the designated purpose of picking up the wounded; although their use was sometimes employed for transporting medical supplies in urgent cases. (Eventually these ambulances were habitually used for the transport of supplies to the brigades and regiments.) Letterman also assigned men to be permanent drivers and he instituted a standard requirement of training for the drivers. Casualties would not be left to fend for themselves on the battlefield, any longer. Next, Letterman devised field hospitals on the division level by removing surgeons from the regimental level and assigning them to division level hospitals. Moreover, each surgeon was tested and was assigned to duty that reflected his skill level. Next, Letterman established field-dressing stations where patients were divided into categories according to the severity of their wounds and were treated accordingly. Letterman's system of organizing patients into those who would live regardless of their wounds and those who would die from their wounds is used to this day and is known as triage.

Battlefields that were not yet introduced to the Letterman system were forced to tolerate inefficient and abusive systems of ambulance use. For example, on August 29, 1862 after the second Battle of Manassas, it was discovered that three thousand wounded were left on the field for three days and six hundred were left for a week. This was discovered since the ambulance drivers who replaced the civilians picked the pockets of the wounded, stole alcohol from the medical supplies and left the injured to die. As if that was not bad enough, the Sanitary Commission could not find a single record of wounded soldiers reaching their destinations after the battle of Bull Run in Virginia. As a response to these and other medical scandals, Letterman's system was gradually adopted by all of the Federal Armies in the field.

The Army of the Potomac was later transferred from the Peninsula to Alexandria, Virginia, where Dr. Letterman discovered that the supplies available were few and deficient. It was soon realized that the rapid and rushed transfer of the army caused supplies and ambulances to be lost or left behind. The medical officers and officers of the Ambulance Corps were also very weary. Yet, the Army of the Potomac marched into Maryland and fought the battle of Antietam under these disadvantageous conditions in September 1862. It was here that the value of Letterman's brilliant new Ambulance Corps system was shown. After that battle, Dr. Letterman decided to make the method of getting medical supplies more efficient. He reduced the amounts of medicines and materials to be carried and reduced the number of wagons used to transport them by half of the previous numbers; thus he was able to convert the transport system into a more compact and functional apparatus of the army. The details of this new arrangement were published on October 4, 1862 and were republished again somewhat later; these new arrangements were held in such high esteem that no further changes were ever found to be necessary.

Additionally, Letterman increased the accountability of the medical department by two principle means: the implementation of medical inspections and the filing of detailed reports by the inspectors responsible for these reviews.

On October 30, 1862 Letterman established field hospitals while the Army of the Potomac was still in Maryland and this system was carefully designed to work with the Ambulance Corps and the means of supply as a whole. It was at the Battle of Fredericksburg that this holistic approach allowed for its first opportunity to prove its worth. Those who were a part of the conflict testified to the system's effectiveness. Surgeon Charles O'Leary, then Medical Director of the Sixth Corps said in his report: "Being appointed Medical Director of the Sixth Corps a few days prior to the Battle of Fredericksburg, December 13, 1862, I had the opportunity of putting in operation the Field-Hospital organization devised by the Medical Director of the Army of the Potomac, and witnessing its beneficial results. Within a very few hours after the positions were designated for the Field Hospitals on December 12, all the necessary appliances were on hand, and the arrangements necessary for the proper care of the wounded were as thorough and complete as I have ever seen in a civil hospital."

During the Battle of Gettysburg in July, Letterman had established a General Hospital and temporary field hospitals wherever there was a source of water and shelter. The buildings used ranged from churches, farm buildings and private homes. Sometimes the only shelter available was trees or a piece of canvas strung between poles. As the battle approached its end, there were approximately 23, 000 wounded from both the Union and Confederate armies that required attention.

THE IRISH BRIGADE

Medical supplies were running low and the hardships of the wounded were equally matched with the exhaustion of the doctors. The site chosen for "Camp Letterman" hospital was on the Wolf Farm a little over a mile from Gettysburg on the York Pike. Although his previous service had exposed him to the turmoil of the battlefield, Letterman could not help but feel devastated by what he called "a vast sea of misery." Although Camp Letterman was primitive by modern standards, it was impressive to the soldiers. Each tent held forty folding cots with mattresses and sheets that proved to be a luxury for soldiers who were used to lying on hard ground upon being wounded.

The practices placed in the Medical Corps by Dr. Letterman were far more recognized in his time that in ours. In one tribute to his contributions in that time we find the following accolade attributed to his efforts,

> "Whatever may be the future of the Army of the Potomac, it has gained a reputation for perfection of organization which will secure it a commanding position among the armies of history.
>
> "But the Medical Department has special claims upon the attention of the country. Without detracting from the merits of the other branches of the Army, we may say that the organization of the Medical Department has attained a degree of perfection that is found in no other army at home or abroad. It will be seen that the reforms were radical and developed, under different heads, a system of operations that covered the whole field of medical service. Its utility consisted in reducing to harmony and concert of action every branch of the medical service, and in placing the right man in the right place; unity and efficiency was the keynote of the reform proposed, and to this every other consideration had to yield. The entire medical staff of the Army became a unit, and moved with the deliberation and precision of a single person. Of the practical value of these improvements we are now able to speak in the most unqualified terms. They have been put to the most rigid test, and have been found in the highest degree practical and effective. The medical staff of no army ever worked in such

perfect harmony and subordination on the battlefield as that of the Army of the Potomac. The battles of South Mountain, Crampton's Gap, Antietam, Fredericksburg, Chancellorsville, and Gettysburg, have placed the most violent strain upon every detail of this organization, whether taken as a whole or in its separate parts, and yet it has never been found wanting. The prompt care of the wounded in these sanguinary battles was never exceeded under similar circumstances.

"The highest attestation of the value of the present organization of the ambulance service of the Army of the Potomac is found in the unanimity with which it has been pressed upon the attention of Congress, and the recent almost unanimous action of that body in extending its provisions to all the armies of the United States. Its system of field hospitals has in the main, also been adopted by the Surgeon-General for all our armies.

"Too much praise cannot be awarded to Dr. Letterman for the patient and intelligent zeal with which he has labored to establish and perfect the present organization of the medical service of the Army of the Potomac. Its conception could only occur to a mind apt in method and organization, and while of comprehensive grasp, yet trained by experience to the study of details. To Dr. Letterman is due the gratitude of the country for his perseverance in effecting these desired reforms."

Medical Times, New York April 30, 1864

Jonathan Letterman received no medals for his efforts or his battlefield innovations, then or later. Nor, were his contributions recognized in a manner deserving of his contributions. His focus was placed primarily upon the task at hand and as such, went unheralded. The present account of Dr. Jonathan Letterman and his role in the Civil War tells the story of an extraordinary person who exhibited a lasting, uncommon dedication coupled with superlative heroic service to his fellow man during a critical time in our country's struggle to redefine its self.

6

LETTERMAN AT SOUTH MOUNTAIN

"My brother died on the field at 1st Bull Run. His body laid unattended for several days and I was later informed by one of his colleagues that he most likely would have lived had he received care in a timely manner. I don't know whether he had suffered, nor what his last thoughts were. But I imagine that he thought about his young wife Hannah and his son Jacob alone and uncared for. Since that time I have often wondered how many other men shared his same fate and whether I could help prevent that. I am here because I want to make a difference in the lives of such men. And I believe that I can."
Medical Orderly *Jack Brody*

Men and women came to the war at different times during its course and at different periods of their lives. Moreover, they did so for a variety of reasons. And that was true of the warriors who fought the battles in many and all places. Letterman entered the war during the crucial second year of conflict, in 1862. At this point many of the combatants in the eastern theatre were placed in position, as well as those that had assumed supportive critical roles that they would play until the end of hostilities in 1865. That is, if they survived.

Among the crucial elements supporting the soldiers' activity were the sources and extent of information to be gleaned from the activities on and off the field of battle. These sources include material offered by networks of spies and informants, as well as, surprisingly, and somewhat unaccountably, newspaper accounts of the battles.

The reports from the battlefield offered the combatants a significant source of information on the size and viability of the fighting forces they and their comrades had encountered, as well as those that they would face. An important contribution to the events that would occur were the roles played by spies, and the networks that served their purpose. Among those reporting to the intelligence community in the North were John Scobell and his associates, among them scouts Ted Knightly and Bog Smith.

As we enter the scene Ted Knight and Bog Smith are in the process of reporting their observations to John Scobell.

"There can be no doubt, I'm afraid. We stopped the Rebels at South Mountain, as you know, but our observations suggest that they are far from returning to Virginia. As always, they haven't given up."

Bog Smith continues the narrative. "As you advised, we stayed behind after the fighting until it was over. And what we saw leads us to believe that more is in store for the Army of the Potomac. And unless I miss my guess, that will happen soon. Whatever General Lee's plan he is not ready to concede to General McClellan. He will continue to fight."

Seeking more detail John asks his couriers, "What exactly did you see? I want to give Mr. Pinkerton a complete report before he reports to General McClellan. And you know that he will insist on learning the numbers of troops that General Lee will send forth against us. So in spite of the number of troops that the Rebels brought with them before the Battles of South Mountain, and the number that did not survive the Battle, a count would be good. While we outnumber them, General McClellan wants to keep two Corps of our boys in reserve as a precaution against any attack on Washington."

As Ted responds, " Well, General Lee sent his Second Corp with General Jackson west, I reckon, to Harpers Ferry to capture the battery stationed there. Aside from his troops, they took a large battery of cannon, probably to attack and to seal off the Federal's escape, with them. As Bog explained the set-up to me, Jackson's expertise in positioning his troops and especially, in placing his guns, should carry him well off. "

As John concludes, "Well, We can't worry about that event. We need to concern ourselves with the large body of remaining troops and their threat to us in this place."

The United States Medical Military Corp, at the initiation of the Civil War in 1861 consisted of 30 surgeons and 83 assistant surgeons. Quickly, this small pool of insufficient numbers of men, moreover, was soon depleted, especially so as men took up different sides in this conflict. Medical personal were not immune to the political persuasions of the time or their affiliations with one side over the other; rather, men with dramatically different points of view or possessing different regional allegiances resigned their medical posts in the regular army and chose instead to follow the destiny of the armies of their chosen preference.

In support of the medical needs concurrent with the war, hundreds of civilian doctors on both sides enlisted to help fill the void created by this paucity of medical personnel. Consequently, these "grafted" doctors, as they were called, untrained in military matters and lacking experience in the care and treatment of men subject to war - affiliated damage that were placed under their care, had to be taught the ways of the military and the unique demands it posed in attending to their charges.

There was at the time an alternative route to becoming a physician. This path entailed, as many professions at the time, an apprenticeship consisting of on-the-job training. A person wanting to become a doctor would work with another medical practitioner, until his chosen mentor felt that his student had learned all that could be taught. At such time the apprentice would be sent out to start his own practice. Needless to way, the practice of the medical arts was limited to men. In spite of the need, there were no schools of nursing, nor the training of women in the caring arts at the onset of the War, at least until some time later, when the need for more attendant care, caused by the large unanticipated number of casualties demanding more care, became evident.

By1860 there were over 100 medical schools and the majority of doctors trained had some type of formal education. Yet, neither learning through direct experience nor formal education could prepare the doctors of the 1840's and 1850's for what was to besiege them the next decade in the Civil War.

The War offered a form of "on the job training" for everyone who worked in the medical field. Clearly, the challenge was substantial and most often, overwhelming; there were few men, or women, for that matter, that had sufficient experience to handle a war like this one. Aside from those few doctors that had served in the Mexican War in the previous decade, experience in the field was lacking; moreover, there was no training or protocol to follow nor venue to assist in that vital task. Neither was there a bank of experience on the part of practitioners nor were there medical traditions available from previous engagements of the present sort among those people that would serve.

The Casualty List

The vast numbers and unusual forms of casualties that occasioned the circumstances of the war both challenged and exceeded our ability to repair ourselves. Of nearly 3 million soldiers who fought in the war, nearly 750,000 would die during this time, among them men who would suffer severe and debilitating wounds, or often be unaccounted for thereafter. Approximately, 400,000 men died from disease, while 218, 000 died in battle or from wounds received during battle. Among the dead, it is believed that the North lost 359, 528 men and the South had lost 258,000 men. These figures, of course, are approximate since more accurate records were neither systematically accumulated nor properly catalogued. Furthermore, in the heat of battle such accounting was difficult and/or haphazard. Many wounded men that were not directly killed in battle would die shortly thereafter and be unaccounted for after the fighting in a given sector terminated.

Acquiring information about the fate of a loved one by his family was often a limited venture; identities were often unavailable as well as not readily attainable. Many dying soldiers expressed the thought that they would hope to be buried at home. Yet, as armies moved over vast distances thousands of fallen soldiers were left in places where they had fallen or were unaccounted for, usually within the place where their bodies often lay unattended or uncared for; unfortunately, knowledge of the identities of these men would not be determined then or beyond. The battlefields were too widely scattered across the country and the men that marched to battle across that "foreign" terrain could only hope to be remembered in memory.

For the lucky few that were rescued, or removed from the fields of conflict adequate care was often neither in place nor assured; most "treatment centers" were informal or haphazard, dependent on the resources of the communities where the fighting had occurred. And in these places the resources of care were limited and inadequate. At Gettysburg a town of 2300 persons at the time of the battle, casualties (killed, missing, or captured) reached over 52,000 at best count.

The hospitals, where they existed, were often in disarray and poorly organized. The many deficiencies that prevailed included insufficient and poor housing, inadequate sanitation and haphazard nutrition, and personnel untrained in the latest medical practices and techniques. Moreover, medical facilities lacked a system of effective accounting of the human costs of this war.

On the evening of May 10th, 1864 – as the Civil War entered its fourth year 26-year-old James Robert Montgomery, a private in the Confederate Signal Corps in Virginia, wrote a letter to his father back home in Camden, Mississippi He was dripping blood on the paper as he wrote. He was suffering from a horrific arm wound he had sustained just a few hours earlier and he would die, as he was aware at the time, shortly. His letter home read as follows:

"Dear Father,

This is my last letter to you. I have been struck by a piece of shell and my right shoulder is horribly mangled and I know death is inevitable. I am very weak but I write to you because I know you would be delighted to read a word from your dying son. I know death is near, that I will die far from home and friends of my early youth but I have friends here too who are kind to me. My friend Fairfax will write you at my request and give you the particulars of my death. My grave will be marked so that you may visit it if you desire to do so. It is

33

optionary with you whether you let my remains rest here or in Mississippi. I would like to rest in the graveyard with my dear mother and brothers but it's a matter of minor importance. Give my love to all my friends. My strength fails me. My horse and my equipments will be left for you. Again, a long farewell to you. May we meet in heaven."
Your dying son, *J. R. Montgomery*

James Montgomery's friend, Fairfax, did write soon thereafter -- forwarding some of his effects -- and assuring his father that he had been conscious to the end, and that he had died at peace with himself and his maker. But it was little consolation. Though the grave had been marked at the time of J.R.'s death, the family was never able to find it, and was thus never able to realize their fond hope of bringing their dead son home.

Obsolete Thinking and Outmoded Solutions

Compounding the problems produced by injuries caused by the war, an understanding of antisepsis in the treatment of the infirmed was just becoming known and accepted in America. That sphere of medicine was influencing the frontiers of medical practice in other places, primary Europe, where the Crimean War (1853-1856) was recently fought, but not in the United States. This lack of knowledge/awareness concerning the role of infection and its prevention in the causes of illness/injury resulted from the infrequent and haphazard employment of basic hygienic practices and sanitary principles. Among common practices, participating doctors in a rush to treat the men placed under their care unknowing of the effects/prevention of contamination would seldom wash their instruments or their hands (or replace and discard used bandages among the men), extending opportunities for infection and disease to occur, while attending to different patients. Moreover, after surgical repair had been performed, their patients were often subject to contagion as they were generally housed in ill-prepared and over crowded quarters, while food and water was often contaminated by fecal material produced by the men and their animals. With overcrowded conditions inherent in such arrangements, and the opportunity for contagion ever present, the spread of infectious diseases was rampant and devastating among those soldiers placed in these circumstances.

The call for changes in practice and procedure was of utmost importance and was recognized as such. However, change could not be rushed.

Innovations in battlefield medical care came about in the early part of the Civil War in response to the horrendous conditions wounded soldiers faced. A classic example was the Battle of Second Manassas (Second Bull Run), a clash in late August 1862 in which Confederate troops defeated Federal forces outside of Washington, D.C. In that engagement more than 22,000 soldiers were killed or wounded in the conflict—nearly 14,000 of them Federal troops. In the aftermath of the battle wounded soldiers were strewn about the battlefield, crying out for water and medical attention. But there was little comfort for those needing care. Left on the open field to fend for themselves, and without necessary attendant care, their shrieks and moans increased as the days wore on. A full week would pass before all of the Union injured were removed from the battlefield and transported to hospitals—grisly proof of the Army's appalling inefficiency in dealing with casualties.

Often, even when a soldier was rescued from the battlefield, a problematic occurrence, his chances of survival depended upon his effective treatment in the hospital where he was taken. However, many proved to be ineffective and worse.

Something had to be done. However, there were few men with the skill and experience to take up the task. Yet, fortunately, there were a few equal to the undertaking. One of these men was Dr. Jonathan Letterman, a physician from Pennsylvania. Over time, he and several others were prepared to face and defeat the challenges posed by wartime medicine.

Miraculously, the outlook for the wounded changed within two weeks after Jonathan Letterman's arrival at the scene of the latest fighting. On September 17, 1862, Federal and Confederate forces clashed again in Sharpsburg, Maryland, at the Battle of Antietam. After 12 hours of fighting, some 23,000 men (over 12,000 Union soldiers and more than 10,000 Confederates) had fallen—the bloodiest single day of combat in the War, and in American history. In stark contrast to the delays experienced at Manassas earlier, however, every injured Union soldier was evacuated from the Antietam battlefield within 24 hours. This change can be attributed to the appointment of Dr. Letterman, the new chief of medical services, physician from Pennsylvania. Dr. Letterman's appointment to this post did not come too soon.

7

THE APPOINTMENT

A s the Surgeon-General reported, the medical situation at this point in the war was perilous. Moreover, he was not alone in either his observations, or his concerns. In his report of prevailing conditions, General George B. McClellan also alluded to the depth of the problem. He noted that, "the nature of the military operations had also unavoidably placed the Medical Department in a very unsatisfactory condition. Supplies had been almost exhausted or necessarily abandoned; hospital tents abandoned or destroyed and the medical officers deficient in numbers or broken down by fatigue."

MAJOR JONATHAN LETTERMAN

At the time of Letterman's appointment the Army of the Potomac had suffered heavy casualties several months earlier at Harrison's Landing on the James River, where the army had retired after McClelland's failed service in the Peninsula Campaign of 1862.

Major Jonathan Letterman

The practices placed in the Medical Corps by Dr. Letterman were far more recognized in his time that in ours. In one tribute, offered him at the time, it was written:

> "Whatever may be the future of the Army of the Potomac, it has gained a reputation for perfection of organization which will secure it a commanding position among the armies of history.
>
> "But the Medical Department has special claims upon the attention of the country. Without detracting from the merits of the other branches of the Army, we may say that the organization of the Medical Department has attained a degree of perfection that is found in no other army at home or abroad. It will be seen that the reforms were radical and developed, under different heads, a system of operations that covered the whole field of medical service. Its utility consisted in reducing to harmony and concert of action every branch of the medical service, and in placing the right man in the right place; unity and efficiency was the keynote of the reform proposed, and to this every other consideration had to yield. The entire medical staff of the Army became a unit, and moved with the deliberation and precision of a single person. Of the practical value of these improvements we are now able to speak in the most unqualified terms. They have been put to the most rigid test, and have been found in the highest degree practical and effective. The medical staff of no army ever worked in such perfect harmony and subordination on the battlefield as that of the Army of the Potomac. The battles of South Mountain, Antietam, Fredericksburg, Chancellorsville, and Gettysburg, have placed the most violent strain upon every detail of this

organization, whether taken as a whole or in its separate parts, and yet it has never been found wanting. The prompt care of the wounded in these sanguinary battles was never exceeded under similar circumstances.

"The highest attestation of the value of the present organization of the ambulance service of the Army of the Potomac is found in the unanimity with which it has been pressed upon the attention of Congress, and the recent almost unanimous action of that body in extending its provisions to all the armies of the United States. Its system of field hospitals has in the main, also been adopted by the Surgeon-General for all our armies.

"Too much praise cannot be awarded to Dr. Letterman for the patient and intelligent zeal with which he has labored to establish and perfect the present organization of the medical service of the Army of the Potomac. Its conception could only occur to a mind apt in method and organization, and while of comprehensive grasp, yet trained by experience to the study of details. To Dr. Letterman is due the gratitude of the country for his perseverance in effecting these desired reforms."
Medical Times, New York on April 30, 1864

This present story offers an account of Dr. Jonathan Letterman, a young doctor, though the application of his medical skills and his dedication to service during a critical time in our country's internal struggle to redefine itself saved thousands of lives and revolutionized the practice of battlefield medicine. During his time in service the nation was unprepared for a sustained civil war; moreover, a major war was neither anticipated nor prophesied by most. But fortunately, Jonathan Letterman and the Corp of medical practitioners he created, over the course of the first years of the War, were able to meet the challenges posed by the conflict.

The initiation of the practice of battlefield medicine stemmed from humble beginnings. The Civil War as it evolved offered a form of "on the job training" for everyone who worked in the medical field. Clearly, the challenge was substantial and most often, overwhelming; there were few men, or women, for that matter, that had sufficient experience to handle the consequences of a war like this one. Aside from those few doctors that had served in the Mexican War in the previous decade, experience in the field was lacking; moreover, there was no training or protocol to follow nor venue to assist in that vital task. Neither was there a bank of experienced men and women nor medical traditions available from previous engagements of the sort present among those people that would serve.

The numbers and forms of casualties that occasioned the circumstances of the war exceeded our ability to repair ourselves. Of nearly 3 million soldiers who fought in the war, over 750,000 would die during this time, including others would suffer severe and debilitating wounds, or likely to be unaccounted for thereafter. Approximately, 400,000 men died from disease, while 218, 000 died in battle or from wounds received during battle. Among the dead, it is believed that the North lost 359, 528 men and the South had lost 258,000 men. These figures, of course, are approximate since more accurate records were neither systematically accumulated nor properly cataloged. Furthermore, in the heat of battle such accounting was difficult and/or haphazard. Many wounded men that were not directly killed in battle would die shortly thereafter and be unaccounted for after the fighting in a given sector terminated.

Adding to the misery, acquiring information about the fate of a loved one by his family was often a limited venture; the identities of the fallen were often unavailable as well as not readily attainable through an accurate count. Moreover, many of these fallen warriors were, because of the circumstances surrounding the battlefields that they fought upon, left in a precarious state thereafter, neither attended to nor saved from the ravishes of the battle.

In spite of these conditions many dying soldiers expressed the hope that they would be buried at home. Yet, as armies moved over vast distances thousands of fallen soldiers were left in places where they had fallen or were unaccounted for, whether by the place where their bodies often lay unattended or cared for; unfortunately, knowledge of the identities of these men would not be determined then or later. The missing in action were neither recognized nor adequately tallied. The battlefields were too widely scattered across the country and the men that marched to do battle across that "foreign" terrain could only hope to be remembered in memory.

8

THE SPIES

As preparations for the first of the battles directly preceding Antietam on September 17, 1862 were put into place a variety of discussions were in process among the officers and men poised to commence battle. On South Mountain the following conversation occurred between Major General Winfield Scoot Hancock and John Scobell, his black spy.

"The death toll of this war is far too great, and there is no doubt in my mind that it will mount over the course of this war. It is a burden that none of us can bear, blue or gray. But we must continue our efforts since we cannot come to terms on how to end the bloodshed. Yet, I think, that in spite of this burden, we can do better. I have a very delicate assignment for you in this regard. I know that it will be appear to be unorthodox and new to you, as it is to me," said the General.

"It is an assignment that I have never asked nor commissioned a man to do. But, if I am right, it is imperative. And you can do it."

"It sounds important," says John in response.

"I believe that it is imperative. I want you to protect a very important man whom can help us endure the cost of the war and limit our sacrifices. He is Dr. Jonathan Letterman our chief medical officer. The good doctor is a remarkable man whose ideas about the practice of battlefield medicine, the treatment of the injured, and protection and well being of the healthy soldier have come due. And his insight into these problems should place us in good stead."

"From my discussions with him, I believe that if given the opportunity his ideas about field medicine, hospitalization and sanitation, as well as some of his ideas about the selection and preparation of our food, we can serve our wounded better, speed their recovery and insure among the provisions we provide the able soldier that our army will be in sound physical condition to pursue and defeat the enemy. It is impossible to predict to what degree Dr. Letterman will effectively address the mortality rate among our soldiers. Nor can we predict the degree to which his wisdom can contribute to the well being of our troops. Yet, he is the man that can most aid our efforts in this regard. And we cannot ignore his value to us. Nor, apparently can others, as well."

"Dr. Letterman's value to us, unfortunately, has not gone unnoticed in the enemy's camp. But there is also recent news to report on this front."

"Our intelligence agents suggest that Dr. Letterman's worth to the Army has not gone unnoticed by the enemy. We also have reason to suspect that they have assembled a small team of people that have been commissioned to interfere with his important duties on our behalf. If our sources are correct, and our assessment of their intent is accurate then they may be ready to kidnap or kill the good doctor, possibly by poison or otherwise, to prevent though this means, the implementation of his medical goals on our behalf. I need not tell you, but we cannot allow this to happen, if we can prevent it. And I believe that through you Mr. Scobell, based upon your past contributions to our cause that we must try to prevent that from happening."

"But, …sir", protests a sputtering John. "I never have been a bodyguard, nor have I been involved in the task of protecting a person from harm. This is beyond my realm of experience. I wouldn't know where to begin. Nor, is it likely that I can, as one man alone, stop a team of determined men from committing so foul a deed. I would need the assistance and expertise of others in order to accomplish this mission. I would also need access to our latest weapons, as well as your complete support when called upon. I could not operate alone. This would be a very different assignment than those I have confronted in the past."

"Well, Mr. Scobell, you are right of course, and your concerns are fully justified. But with the proper people as aides, with our full intelligence network at your disposal, and with the most effective and advanced weapons that we can provide, I think that you can do it. Your experience in organizing and directing men and your ability to lead them should allow you to complete this thankless task. And I know, in spite of the difficulty involved in this assignment that you can do it."

"I will allow you to select four men now under your direction to assist you, and they would be under your command as well. Moreover, I can offer you the assistance of as many others, as may be needed. Aside from outfitting you with weapons, good horses, maps, and providing you information regarding our adversaries we can offer you any additional support available to this command, as it is needed. This task, as you might surmise demands the very best effort that we can muster. I will put forth and endeavor to provide you and your men as much support that I am able."

"For weapons we can start by providing you an assortment of our latest firearms as well as explosives to assist you in your charge. For starters I will have our quartermaster issue you six colt revolvers and 1800 rounds of ammunition.* That amount should suffice for practice and performance, as the need arises.

"'You will also need good rifled muskets and we have several types including the Henry, with appropriate rounds of ammunition. There will be no question regarding your ability in providing yourselves the personal protection that may be required, nor for the protection of Dr. Letterman, as circumstances warrant. And we can add to these safeguards as you might suggest.

Editor's Note: *Pistols at the time of the War were single-action weapons based on the Colt design. Soldiers armed with them often carried pre-loaded cylinders to swap out of their weapons rather than having to reload the gun in the midst of battle. Soldiers and officers on both sides often carried a pistol as a secondary weapon. Union soldiers typically carried Colt revolvers. Patented in 1835 by Samuel Colt, the revolver featured a spinning cylinder that allowed the soldier to fire five or six shots instead of the one or two fired by earlier pistols.

"I know that this is new work for you Mr. Scobell. And I know that it is nasty task, as well as risky and dangerous. And that it may take some time before we can wipe out this viper's nest that resides within our midst and that threaten us through the efforts of Dr. Letterman. But it is doable. And I will offer you complete cooperation and rely upon your discretion, as always. After giving this plan a lot of thought, I have come to the conclusion that you are uniquely suited to the task and I sincerely hope that you will accept the job."

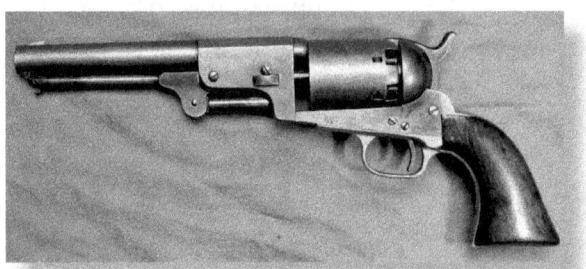

SAMUEL COLT REVOLVER

Confederate Spies in Washington

The heart of the spy network for the South was established within northern territory, precariously located 60 miles south of the Mason-Dixon Line, in Washington D.C. Prior to the war the national capital was a hotbed of southern sympathizers and those aiding its cause. Among them, was Virginia's Governor John Letcher a former congressman familiar with the city. He had used his knowledge to set up an emerging spy network in late April 1861, within the short period of time during which Tennessee seceded from the Union and as it was about to join the Confederacy.

Two of Letcher's most prominent early recruits were Thomas Jordan, a West Point graduate stationed in Washington before the war, and Rose O'Neal Greenhow, an openly pro-South widow and socialite, friendly with a number of northern politicians, including Secretary of State William Seward and Massachusetts Senator Henry Wilson.

In July 1861, Greenhow sent coded reports across the Potomac to Jordan (now a volunteer in the Virginia militia) concerning the planned Federal invasion. One of her couriers, a young woman named Bettie

Duvall, dressed as a farm girl in order to pass Union sentinels on the Chain Bridge leaving Washington, then rode at high speed to Fairfax Courthouse in Virginia to deliver her message to Confederate officers stationed there. Confederate General P.G.T. Beauregard later credited the information received from Greenhow with helping his rebel army win a surprise victory in the First Battle of Bull Run (Manassas) on July 21, 1861.

Confederate Signal Corps and Secret Service Bureau

Although less organized that its Northern counterpart, the Confederate Signal Corps, which operated the semaphore system used for communicating vital information between armies on the field, also set up a covert intelligence operation known as the Secret Service Bureau. Headed by William Norris, a former Baltimore lawyer who also served as chief signal officer for the Confederacy, the bureau managed the so-called "Secret Line," an ever-changing system of couriers used to get information from Washington across the Potomac and Rappahannock Rivers to confederate officials in Richmond. The Secret Service Bureau also handled the passing of coded messages from Richmond to confederate agents in the North, Canada and Europe.

A number of Confederate soldiers, especially cavalrymen, also acted as spies or "scouts" for the rebel cause. Among the most famous were John Singleton Mosby, known as the "Gray Ghost," who led guerrilla warfare attacks in western Virginia through the latter years of the war, as well as J.E.B. Stuart, the celebrated, flamboyant cavalry officer whom General Robert E. Lee called "the eyes of the army."

Union Spies: Allan Pinkerton's Secret Service

Allan Pinkerton, the founder of his own detective agency in Chicago, had collected intelligence for Union General George B. McClellan during the first months of the Civil War, while McClellan led the Department of Ohio. When President Abraham Lincoln summoned McClellan to Washington late that summer, the general put the detective in charge of intelligence for his Army of the Potomac. Thereupon, Pinkerton set up the first Union espionage operation in mid-1861. Calling himself E. J. Allen, Pinkerton built a counterintelligence network in Washington and sent undercover agents to ingratiate themselves in the Confederate capital of Richmond. Unfortunately, Pinkerton's intelligence reports from the field during McClellan's 1862 Peninsula Campaign and thereafter consistently miscalculated Confederate troop numbers at twice or three times their actual strength, fueling McClellan's field timidity and his repeated calls for more reinforcements. It was an unfortunate confluence of influences, feeding into the general's concern for his men and his undue caution and strong reluctance to act on the battlefield.

Though he called his operation the U.S. Secret Service, Pinkerton actually worked only for McClellan. Union military intelligence was still decentralized at the time, as generals (and even President Lincoln) employed their own agents to seek out information and report back to them.

Another prominent Union intelligence officer was Lafayette C. Baker, who worked for the former Union General-in-Chief Winfield Scott and later for Secretary of War Edwin Stanton. The brave but ruthless Baker was notorious for rounding up Washingtonians suspected of having southern sympathies; he later directed the manhunt for John Wilkes Booth, the actor and Confederate sympathizer who shot and killed President Lincoln at Ford's Theatre in April 1865.

Prominent Civil War Spies

Thanks to her success, Rose O'Neal Greenhow was one of the first Confederate spies targeted by Allan Pinkerton. Shortly after the southern victory in the First Battle of Bull Run, Pinkerton put Greenhow under surveillance and subsequently arrested her. Imprisoned in the Old Capitol Prison, she was released in June 1862 and sent to Richmond.

Belle Boyd, another famous southern belle-turned-Confederate spy, helped smuggle intelligence to General Stonewall Jackson during his Shenandoah Valley campaign in 1862. Like the Confederacy, the Union also made use of female spies: Richmond's Elizabeth Van Lew, known as "Crazy Bett," risked her life running an espionage operation out of her family's farm, while Sarah Emma Edmonds disguised herself as a black slave to enter Confederate camps in Virginia.

The British-born Timothy Webster, a former police officer in New York City, became the Civil War's first double agent. Sent by Pinkerton to Richmond, Webster pretended to be a courier on the Secret Line, and managed to gain the trust of Judah P. Benjamin, the Confederate secretary of war (later Secretary of State). Benjamin sent Webster to deliver documents to secessionists in Baltimore, which Webster promptly passed on to Pinkerton and his staff. Webster was eventually arrested, tried as a spy, and sentenced to death. Though Lincoln sent President Jefferson Davis a message threatening to hang captured Confederate spies if Webster were executed, the death sentence was carried out in late April 1862.

9

THE PLOT TO KILL LETTERMAN

Our story continues, somewhat sooner than John would have anticipated. Moreover, the stage is now set within a fortnight. In the case
at issue a cast of actors has been assembled and the scheme itself expanded from the mundane to the exotic.

It was a bizarre scheme, certainly beyond the realm of any ordinary
person's imagination or of his or her limited daring; by proper accounts
it was a plan devised by a person uniquely suited to complete its intent.
Moreover, in spite of its distinctiveness and extraordinary boldness, it
appeared more than simply plausible, but likely to succeed, if allowed.

The Smallpox Plot

The use of contaminates as a weapon of combat predates the Civil War
by several millennia. From dead animals to feces tossed over medieval city
walls, to the exchange of contaminated clothing of those that succumbed
to various deadly infestations, disease has played an active role in the most
deadly of wars over the centuries. And that was the case in this war, as well.

During the course of the Civil War many prominent people were
involved in the promotion of victory by any means. Among them, was
Dr. Luke Blackburn, a Southern sympathizer from Kentucky, who tasked
himself with a unique mission in spreading contaminants of his own
creation. As a fellow physician Blackburn knew of the varied innovations

in the medical practices of his day, including the work of Dr. Letterman. He quickly recognized the value of this work. From reports derived from the field, he had learned that Letterman was in the process of developing unique approaches to the care and treatment of soldiers. Above all, he knew that it would offer the Federals considerable relief from the casualties suffered on and off the battlefield, an outcome of significant importance to warrant his attention and expertise; he needed to act if he was to neutralize the effects of Letterman's improvements in medical care.

He feared that Letterman's knowledge of the treatment of injuries and prevention of disease, coupled with his willingness to act upon his own recommendations would strengthen the forces of the North to the comparable detriment of the South. Seizing upon his knowledge of disease Blackburn was determined to blunt the advantages offered to the North by Jonathan Letterman.

The plot Blackburn developed, with the approval of the Confederacy's Secretary of War Judah Benjamin was based on his medical knowledge as a specialist in the treatment of infectious diseases. Its ingenuity lay in the fact that it was a plan of rather simple design, one easily executed if properly directed, and well suited to those that would be called upon to implement it. At its core it would involve a few trusted men that would gather and sell clothing containing the smallpox virus to unsuspecting persons who would then, unwittingly, distribute them to members of the Army of the Potomac. In such fashion it was hoped to produce massive casualties among members of the Army of the Potomac. And it was unique insofar as it was properly reasoned that Dr. Letterman, as the principal sanitary officer and supervisor of clothing purchases, would be an unknowing instrument in the distribution of such items while, he too would be exposed to the deadly toxins emitted by such articles.*

Blackburn's plan was neither original nor new; somewhat earlier in the history of the republic, a similar proposal in an attempt to infect Indians with smallpox during the colonial wars was adopted. It had failed then. However, it was now being held in reserve with renewed, if questionable promise, during these tumultuous times.

*Reference to this earlier appearing incident appeared in a book familiar to Blackburn. It concerned a young U.S. officer, Charles W. Randall who was a lieutenant in the Seventeenth Vermont. According to this report Randall's health became permanently impaired by smallpox, which it was believed he contracted from infected clothing, reputedly undergarments, purchased in Washington, D.C. at a store having thereafter come under suspicion as a place of consignment. Unfortunately, Lieutenant Charles W. Randall did not escape the disease sustained under these conditions. In time, the disease destroyed his blood, and shortly after the war ended he died of consumption.

Blackburn's career in espionage specializing, as an undercover agent directing his activities against the North, was well founded. He was no novice to the task; previously, he participated in a variety of schemes of a similar nature over an extended period of time.

Blackburn was also uniquely suited to instigate the current plot. Aside from his medical credentials he was an ardent secessionist and radical. While he was too old to serve actively in the military at the time of the Civil War, a disappointment to him no doubt, he directed or participated in undercover activities for some time throughout the War. In the early days of the rebellion he acted as a civilian agent for the governments of Kentucky and Mississippi. By 1863, he was aiding Confederate blockade runners in Canada and in 1864, he traveled to Bermuda to help combat a yellow fever outbreak that threatened Confederate blockade running operations there.

Shortly after the War's end, a Confederate double agent accused him of having carried out a plot to start a yellow fever epidemic in the Northern United States that would have hampered the Union war effort. Like his small pox scheme Blackburn was accused of collecting linens and garments used by yellow fever patients and smuggling them into the Northern states to be sold to unwary customers. The evidence against Blackburn, at the time was considerable. However, much of it was circumstantial. Moreover, the sources of this information, as it turned out, were of questionable character.

The evidence for conviction was good but not overwhelming. Consequently, as luck would have it, Blackburn was tried but acquitted by a Toronto court in Canada for his role in the plot; nevertheless, public sentiment was decidedly against him throughout much of the United

States. His plot, moreover, would not succeed in spite of Blackburn's efforts. It is interesting to note that sometime later Blackburn's plan was shown destined to fail; in 1900, long after the Civil War, Dr. Walter Reed discovered that Yellow Fever, in spite of beliefs held at the time, is typically spread by mosquitoes not by other forms of contact.

The Plot Emerges

Approximately one month before the meeting between General Winfield Scott Hancock and John Scobell another gathering was underway in Washington, D. C. It was being held in a secluded room of a boarding house known to accommodate southern sympathizers, among them Dr. Luke Blackburn. The doctor had organized this meeting with the preapproval, as well as the financial support of the Davis administration under the guidance of Judah Benjamin, one of the president's closest political advisors. While attempting to maintain secrecy, however, Blackburn's reputation, nevertheless, followed him. Suspicion soon fell upon him. His well-known connections and nefarious activities both unknown to him and his followers at the time were being closely monitored by Federal agents. These men would be assigned to watch and report on Blackburn's activities.

DR. LUKE P. BLACKBURN

On the night of the gathering, Blackburn laid out his plan to his associates. Two were secretly assigned to buy linens and impregnated clothing from hospitals that specialized in the care and treatment of soldiers stricken with smallpox. Normally, as a precaution, such material was removed and burned, as ordered, rather than being washed and sold as surplus goods. However, Blackburn's agents had encountered a corrupt civilian chamber attendant, one of many civilians seeking profit, that were employed in the hospitals, who was desperate for money and saw no problem selling soiled linen that was likely to be disposed of in the near future anyway. To the attendant's wife, he assigned the task of washing and bagging the remnants of the linen and clothing seized from beds and closets where they were stored prior to disposal. These articles, he instructed her, should be placed in sealed bags, stored for a limited time, and then scheduled to be picked up and delivered to prospective buyers by himself, under cover of darkness.

Blackburn's agents, having made contact with the hospital attendant, were instructed to help gather the materials for sale to be distributed and sold at half the market price to the very same federal army, the original source of these materials, as surplus with the profit from sales being distributed among those participating in the scheme.

It was a good plan, certainly one of creative imagination, likely not to be noticed nor questioned and reflective of its time and circumstance. But, fortunately it would fail because of the suspicions and the inadvertent intervention of one woman.

Hannah Ropes - Civil War Nurse

She was an extraordinary woman by any standards, a person of high standards and an exceptional level of accountability. And by her actions she saved lives. In her book, Hospital Sketches, Louisa May Alcott described Hannah Ropes' actions [the matron] as casualties arrived from the Battle of Fredericksburg (December 13, 1862):

"In they came, some on stretchers, some in men's arms, some feebly staggering along propped on rude crutches, and one lay stark and still with covered face, as a comrade gave his name to be recorded before they carried him away to the dead house. All was hurry and confusion; the hall was full of these wrecks of humanity, for the most exhausted could not reach a bed till duly ticketed and registered; the walls were lined with rows of such as could sit, the floor covered with the more disabled, the steps and doorways filled with helpers and lookers on; the sound of many feet and voices made that usually quiet hour as noisy as noon; and, in the midst of it all, the matron's (Hannah Ropes) motherly face brought more comfort to many a poor soul, than the cordial draughts she administered, or the cheery words that welcomed all, making of the hospital a home."

Hannah was a humanitarian with strong moralistic inclinations and little tolerance for the mistreatment of her charges, those suffering from the misfortunes of war. As she herself wrote of the events she bore witness:

"The sight of several stretchers, each with its legless, armless, or desperately wounded occupant, entering my ward, admonished me that I was there to work, not to wonder or weep; so I corked up my feelings... The merciful magic of ether was not [used today]. It's all very well to talk of the patience of woman, but the endurance of these men... their fortitude seemed contagious, though I often longed to groan for them, while the bed shook with the irrepressible tremor of their tortured bodies. Forty beds were prepared, many already tenanted by tired men who fell down anywhere, and drowsed till the smell of food roused them.

Round the great stove was gathered the dreariest group I ever saw—ragged, gaunt and pale, mud to the knees, with bloody bandages untouched since put on days before; many bundled up in blankets, coats being lost or useless; and all wearing that disheartened look which proclaimed defeat, more plainly than any telegram of the [General Ambrose] Burnside blunder. I pitied them so much, I dared not speak to them, though, remembering all they had been through since the route at Fredericksburg, I yearned to serve the dreariest of them all."

And later in the day she finishes with this message:

"At five o'clock a great bell rang... The new comers woke at the sound; and I presently discovered that it took a very bad wound to incapacitate the defenders of the faith for the consumption of their rations; the amount that some of them sequestered was amazing; but when I suggested the probability of a famine hereafter, to the matron, that motherly lady cried out: 'Bless their hearts, why shouldn't they eat? It's their only amusement; so fill every one, and, if there's not enough ready to-night, I'll lend my share to the Lord by giving it to the boys."

Editor's Note: And, whipping up her coffee-pot and plate of toast, she gladdened the eyes and stomachs of two or three dissatisfied heroes, by serving them with a liberal hand; and I haven't the slightest doubt that, having cast her bread upon the waters, it came back buttered, as another large-hearted old lady was wont to say.

Hannah was also an activist; she was not content to simply be part of the momentous events of her time. She was a dedicated humanitarian, as well. She needed to improve upon things, particularly the treatment and condition of the men in her care. And she did so, often to the surprise of those in positions of power.

She actively criticized the appalling conditions at the hospital, on the lack of sanitation and harmful circumstances around her and on the indifference and even cruel treatment of caregivers toward the soldiers by those charged with such responsibilities. She believed that every soldier deserved healthy surroundings, good food and humanitarian treatment, and she never hesitated to go to the top to obtain such creature comforts.

She was pro-active in making change, which meant butting heads with the military establishment and often the disdain and lack of respect by the physicians who resented the presence of women in the hospitals. But Hannah was a well-spoken woman and was not afraid to stand up to her male supervisors. As head matron she was appalled by the conduct of some of the doctors, and that of many of the surgeons; she also rejected the capricious selectivity she found among the surgeons in choosing which patients to treat and which to ignore. Many of the surgeons that she encountered seemed to only want to help those soldiers who appeared likely to survive, neglecting the care and treatment of those that would not. She was also displeased with many of the poorly trained attending personnel that she encountered, many of whom saw hospital care as an opportunity to enrich themselves through devious and immoral practices at the expense of their charges.

In her diary she wrote of the following encounter on October 11, 1863:

> "The new steward has been to my room to talk about the washing. I show him how badly the clothes are washed. I did not know till after he left the room that he kept back a part of the ration of soap. I think the hardest thing for me to comprehend is such meanness.
>
> He seemed to be trying to buy me, and I was involuntarily getting at the quality of his mind; he spoke rather contemptuously of the privates. I fired up at that and told him, They were really the heroes of the war and that there were privates in the house who were independent so far as money makes men so, and they knew what their rights were as privates. He said, with a sneer, he was not the benevolent kind, and that he was here to make all the money he could out of the hospital, adding triumphantly that power was in his hands, that he had sent away three loads of their clothes now!"

In a later entry, one most pertinent to our story, she continued her tale of instances of greed and malfeasance in the exploitation of hospitalized soldiers. However, in this case, it was a story with a different ending. With reference to the particular steward she identified in the above incident, she records the following surprising encounter in a letter to her daughter Alice. Hannah writes regarding the behavior of this steward tending to one of the soldiers:

> "The steward and I have not finished up yet—he struck a boy with a chisel and put him in the guard house. "He did not know that even women could send telegraphs, and was rather taken aback the next morning when the father of the boy arrived in Philadelphia before breakfast and demanded the boy! Before eight a man came in and asked for what he put him in here for. The steward said with an oath, "It's none of your business". The man passed up the stairs into the ward and saw the boy's room, asked for the name of the steward and head surgeon, wrote it down with a pencil, and thrusting it

into his pocket his shabby coat fell open, revealing a General's strap! Last night General Banks arrived; he asked for me –the cup of chagrin to the steward seemed full!"

The attendant's surprising rebuke was a well-deserved response to his unacceptable behavior.

Nurse Ropes was a keen observer of the behavior of those around her irrespective of status or rank; and her accounts of those she observed were not limited to civilian personnel. Moreover, Hannah's observations of hospital practices and personnel were reported to her superiors among them ranking members of the president's cabinet. In one such report, Hannah Ropes detailed the following observation:

> "The head surgeon was also a new man, tall, stiff, thin, light hair, whitey blue eye, and whitey yellow complexion, glasses on eyes, and a way of looking out at the end of his glasses at you surreptitiously, if I may use such a big word. He was young and I took to him. He was ignorant of hospital routine; ignorant of life outside of the practice of a country town, in an interior state, a weak man with good intentions, but puffer up with the gilding on his shoulder straps. If he had not been weak, and it had been my style to make a joke at the expense of others, there was a fine chance here; but he was safe at my hands, for he was weak, and I am strong in the knowledge at lest which comes with age. And it is likely that in some way even this man, made giddy with an epaulet, will learn that God has made the private and officer of one equality, so far as the moral treatment of each other is concerned."

Others in authority were less amused by such observations. Repulsed by reports of such malfeasance, Secretary of War Edwin Stanton personally took action against officers and stewards that Hannah found to be slovenly and incompetent. In December 1862, Hannah had a surgeon at Union Hotel Hospital arrested for graft in which he used his official position for profit; he was selling food and clothing meant for hospital patients while pocketing his ill gotten gains.

Budding author Louisa May Alcott arrived at the hospital to work as a nurse soon after her 30th birthday. She was just in time to meet the wounded pouring in from the Battle of Fredericksburg (December 13, 1862) whose plight she recorded. She and Hannah were typical of the women who served in the hospitals.

In Hannah's final diary entry in December 1862, writing in the third person, she described the last words of one soldier passing on from this life to the next: "'Thank you, madam, I must be marching on.' So said Lewie as he passed away. Sitting on one side of him was his nurse, Miss Alcott, on the other side the matron [Hannah]... There was in the man such a calm consciousness of life, such repose in its secure strength... The matron is left alone when the breath ceases."

*In January 1863, Hannah Ropes and Louisa May Alcott contracted typhoid pneumonia - the major killer of wounded soldiers at the time. Miss Alcott hovered between life and death, and watched as Hannah Ropes died of the disease on January 20, 1863, at the age of 53. The day after Hannah died, Alcott returned home to Concord, Massachusetts, where she suffered a long recovery.

A Conversation and a Resolution to the Plot.

The smallpox scheme was now set in motion. Fortunately, the plan was uncovered before being initiated and a counter offensive could now be pursued. Major General Hancock fulfilled his pledge of assistance. His agents, upon learning the details of the plan, were able to report to the general, who in turn, would pass on the information to his trusted assistant. Within a day his aide-d-camp held a meeting with John and his men to inform them of the news that General Banks had provided him. He had also been diligent in organizing a group of men, under John's direction that would detain and arrest the conspirators.

Under close examination by General Bank's counsel, evidence of the scheme was revealed without difficulty; apparently, the good doctor had relied upon a coterie of associates that, more interested in self-preservation than personal sacrifice were willing to cooperate with their interrogators. Upon being faced with evidence of the scheme, they appeared more interested in personally absolving themselves from the plot, in order to avoid punishment, sure to be swift and certain to be administrated, than in their concern for the protection of their fellow co-conspirators. These underlings were caught and they knew it. Their only available course of action was to plead compliance to the scheme under extraordinary pressure and extricate themselves as best as they could by exposing their co-conspirators in hopes of receiving lighter punishment.

Subsequently, the principal players recruited by Blackburn were arrested and taken into custody. All of the participants were apprehended, that is, with one notable exception. Unfortunately, the arrests did not include Dr. Blackburn. Again he eluded the law and his mischief would continue.

For John, however, the termination of the scheme without harm was cause for celebration; Dr. Letterman was safe and able to pursue his reforms. Damage control was applied and it was successful. John knew that his interactions with Dr. Blackburn would not end here. However, knowing of the extremes that Blackburn would apply, if given the opportunity, he would be prepared for their next encounter.*

Editor's Note: *Dr. Blackburn's career as a saboteur, and as a man inclined to use his special knowledge of disease warfare against his perceived enemies in the north did not end with John's intervention in the case of Dr. Letterman. By 1864, the war effort was not going well for the Confederate Army. In response, free to unleash his evil schemes once more Blackburn hatched a plan that was to make macabre use of his area of expertise. In April of that year, he learned that there was a fresh yellow fever outbreak in Bermuda. In response he set sail for the island offering his services as a volunteer to help eradicate the epidemic. However, while still maintaining his initial objective, over the next several months and two trips to the disease-plagued isle, Blackburn set about the grotesque task of secretly collecting his patients' blood-, vomit-, and feces-stained dressings, blankets and clothes. He then packed them into trunks and sent them to Halifax, Canada.

Once the outbreak had passed, Blackburn returned to Canada and arranged to smuggle the trunks south across the border and deliver them to cities like Washington, D.C. where he believed they would cause widespread infection once opened. Legend has it that one particularly fetid trunk was supposed to be delivered to President Lincoln. Blackburn himself proudly referred to his actions as "an infallible plan directed against the masses of Northern people solely to create death." It certainly was a far cry from his physician's oath of "First, Do no harm..."

10

FREDERICKSBURG

At General Robert E. Lee's Headquarters on Maryre's Heights:

The Battle of Fredericksburg did not go as planned. The strategy guiding Burnside's attack would be in the massing of Federal troops in an effort to sweep away all and any opposition in their path. But, while that would guarantee the Federals the benefit of employing a larger contingent of men and material to their advantage, it would be nullified by poor coordination, a poor choice of battlefield, and perhaps most importantly, delays in mounting such an attack. By wanting all of his troops to cross the Rappahannock (the river bordering Fredericksburg) in mass, Burnside's attack afforded Lee the time to mount his response in the proper placement of troops opposing the Army of the Potomac. Burnside's fear in dividing his army, which made it more vulnerable to the enemy, was a common one. However, as Lee had learned (especially, at Antietam), it could be done effectively and with the right Corps commanders. And Longstreet and Jackson, his Corps commanders were ideal officers in that scenario.

The delay in the arrival of the pontoons to be employed in the crossing of the river would play heavily on the outcome of the battle. However, a reluctance to engage in battle was evident elsewhere, as well. These factors were well known to General Lee. Longstreet's spies provided him data to that effect.

On December 11, the federal pontoons were finally laid. However, the day started badly for Burnside's army. For the Federal forces, it began with an unanticipated confrontation. What originally had been planned as an orderly march into Fredericksburg, a city designated by Burnside to serve as a base of operations for an advance on Richmond, became the Battle of Fredericksburg. It was a battle fought from a strategically poor vantage point, amidst poor weather, poor reconnaissance, and as it appeared to many, poor leadership. And it was all of these, and more.

The Battle of Fredericksburg was fought in December of 1862. It was a terrible, misguided approach to the war leaving thousands of federal troops dead and wounded on the battlefield. Many wrote of this ordeal, including Colonel Joshua Lawrence Chamberlain of the 20th Maine. Chamberlain and his troops spent the night of the battle caught between the lines, in this case on the battlefield amidst their dead and wounded. In this place they huddled among their dead and wounded for warmth and protection.

Prequel to the Battle of Fredericksburg

Preceding the battle we find John Scobell and Major Jonathan Letterman on the top of Stafford Hill from their vantage point contemplating the scene of the battle that will occur below them several hours before its initiation.

"This is probably the best view we will have of Fredericksburg", says John Scobell.

He and Dr. Letterman have advanced to the top of Stafford Heights where Major General Ambrose Burnside has positioned his big guns.

John says to his companion, as he points out toward the river and the town behind it, "Just beyond the town, on the other side of the Rappahannock is where the enemy is lodged. Several days ago, I received word that they are stacked up at the wall beyond the field and on the hills behind it. Unfortunately, from our forays into their ranks, we know that they are well dug in and are likely to stay that way. That is, until we make our attack. Then, I'm afraid, they will come out and there will be all hell to pay. General Lee has once again outwitted his Federal adversary."

"But what about these big guns that General Burnside has assembled here?," says the doctor as he turns and points to the massive assembly of cannons occupying Stafford Heights. "Won't they have any effect?"

"I'm afraid that these big guns that we have assembled here will have only a limited result in dislodging them. The trajectory of these guns, while extensive, is limited. Our shells can reach the town, but not the field nor the stonewall beyond the field where the Confederates have assembled with their sharp shooters. And, as I reported to Major General Hancock, General Lee has placed the majority of his troops, including Jackson and Longstreet's Corps, with their cannons on the hill beyond them."

"But …", protests Dr. Letterman, "our men will be slaughtered, even if they advance beyond the stone wall. And that would appear to be unlikely. This seems like a reckless plan, doomed to failure."

"I would agree, as would General Hancock," says John. "And as he has expressed to General Burnside, I might add. That is why I asked you, after consulting with him, to accompany me up here, to get a first hand look at the obstacles facing our troops. And, to alert you to the onslaught that is likely to occur at this place. I'm afraid that you and your doctors will be sorely tested during and after the battle. But you and your system will do well. As before, at Antietam, as we know, you were excellent in your response to the medical crises that occurred there. At least, pray to God that the good work that you and your colleagues and staff have invested in this army will come to good harbor."

"Thank you," says Major Letterman. "But, I don't understand how we got into this situation. We arrived here before Lee and his troops and we could have taken the first initiative, by taking Marye's Heights first. That would have prevented them from seizing the high ground. But we didn't. I just don't understand how that happened."

"I agree. This misfortune could have been prevented. Our plan was a good one", declares John "but we failed to set it off in a timely manner. That delay, unfortunately, will cost us severely. We were unwilling to show the flexibility that General Lee has often demonstrated. It is no wonder that he is often referred to as the "Fox.""

"General Burnside, however, insisted that we wait for the pontoons to arrive, whose request was mishandled, and which, as you know were delayed. He hoped to cross the river as one massive, impenetrable contingent that would sweep across and over the Army of North Virginia. That strategy will fail here as it has in the past. Unfortunately, this Army, although we have frequently outnumbered the Rebels has been plagued from the beginning by that form of reasoning. All of our commanding officers have assumed that the boys in gray were less committed than our troops to this fight and countered on superior numbers and material alone, rather than strategy to take the day. I'm afraid that it is likely that this pattern will continue until we recognize that our adversary is as committed to his cause, as we are to ours. Their generals know this well and exploit it; they use their men to better advantage than we do."

There were many stories each detailing the events that took place at Fredericksburg. One account of the Battle by Colonel Joshua L. Chamberlain is offered here:

A Bivouac with the Dead from

My Story of Fredericksburg
by Joshua Lawrence Chamberlain

"It was a cold night. Bitter, raw north winds swept the stark slopes. The men, heated by their energetic and exciting work, felt keenly the chilling change. Many of them had neither overcoat nor blanket, having left them with the discarded knapsacks. They roamed about to find some garment not needed by the dead. Mounted officers all lacked outer covering. This had gone back with the horses, strapped to the saddles. So we joined the uncanny quest. Necessity compels strange uses. For myself it seemed best to bestow my body between two dead men among the many left there by earlier assaults, and to draw another crosswise for a pillow out of the trampled, blood-soaked sod, pulling the flap of his coat over my face to fend off the chilling winds, and, still more chilling, the deep, many-voiced moan that overspread the field. It was heart-rending; it could not be borne. I rose at midnight from my unearthly bivouac, and taking our adjutant for companion went forth to see what we could do for these forsaken sufferers.

"The deep sound led us to our right and rear, where the fiercest of the fight had held brave spirits too long. As we advanced over that stricken field, the grave, conglomerate monotone resolved itself into its diverse, several elements: some breathing inarticulate agony; some dear home names; some begging for a drop of water; some for a caring word; some praying God for strength to bear; some for life; some for quick death. We did what we could, but how little it was on a field so boundless for feeble human reach! Our best was but to search the canteens of the dead for a draft of water for the dying; or to ease the posture of a broken limb; or to compress a severed artery of fast-ebbing life that might perhaps be so saved, with what little skill we had been taught by our surgeons early in learning the tactics of saving as well as of destroying men. It was a place and time for farewells. Many a word was taken for faraway homes that otherwise might never have had one token from the field of the lost. It was something even to let the passing spirit know that its worth was not forgotten here. Wearied with the sense of our own insufficiency, it was a relief at last to see through the murk the dusky forms of ghostly ambulances gliding up on the far edge of the field, pausing here and there to gather up its precious freight, and the low-hovering, half-covered lantern, or blue gleam of a lighted match, held close over a brave, calm face to know whether it were of the living or the dead.

"We had taken bearings to lead us back to our place before the stone wall. There were wounded men lying there also, who had not lacked care. But it was interesting to observe how unmurmuring they were. That old New England habit so reluctant of emotional expression, so prompt to speak conviction, so reticent as to the sensibilities—held perhaps as something intimate and sacred—that habit of the blood had its corollary or after-glow in this reticence of complaint or murmur under the fearful sufferings and mortal anguish of the battlefield. Yet never have I seen such tenderness as brave men show to comrades when direst need befalls. I trust I show no lack of reverence for gracious spirits nor wrong to grateful memories, when confessing that this tenderness of the stern and strong recalls the Scripture phrase, "passing the love of women."

In additional to official reports filed after the battle of Fredericksburg, a number of other accounts were offered.

The Reverse at Fredericksburg
from *Harper's Weekly,* December 27, 1862

> We have again to report a disastrous reverse to our arms. Defeated with great slaughter in the battle of 13th, General Burnside has now withdrawn the army of the Potomac to the north side of the Rappahannock, where the people congratulate themselves that it is at least in safety. And now, who is responsible for this terrible repulse?
>
> General Burnside was appointed to the command of the army of the Potomac on 9th November, and began at once to prepare to shift the base and line of march of his army toward Fredericksburg. In view of such a movement General McClellan had, before his removal, suggested the propriety of rebuilding and occupying the railroad from Aquia Creek to Falmouth; but, for some reason not apparent, the War Department had not acted upon the suggestion. About 12th November General Burnside notified the Department that he would arrive at Fredericksburg in about a week, and that pontoons must be there by that time, in order to enable him to cross and occupy the hills on the south side of the river."

Burnside's plan was to cross the Rappahannock River at Fredericksburg in mid-November and race to the Confederate capital of Richmond before Lee's army could stop him. It was a feasible plan approved at the highest levels of the federal government. However, it would not be implemented in sufficient time to succeed. Bureaucratic delays prevented Burnside from receiving the necessary pontoon bridges in time to cross the Rappahannock River. As these plans became know, Lee moved his army to block the crossings. When the Union army was finally able to build its bridges and cross under fire, urban combat in the city resulted on December 11–12. Meanwhile, Union troops prepared to assault Confederate defensive positions south of the city and on a strongly fortified ridge just west of the city known as Marye's Heights.

On December 13, having divided his troops into three corps or "grand divisions", Burnside ordered his Corp commanders to move. The first of these led by Maj. Gen. William B. Franklin was able to pierce the first defensive line of Confederate Lieutenant General Stonewall Jackson to the south. However, after much effort his troops were finally repulsed. Burnside then ordered the grand divisions of Maj. Generals Edwin V. Sumner and Joseph Hooker to make multiple frontal assaults against Lt. General James Longstreet's position on Marye's Heights. All of these assaults were also repulsed with heavy losses. Finally, on December 15, Burnside withdrew his army, ending another failed Union campaign in the Eastern Theater.

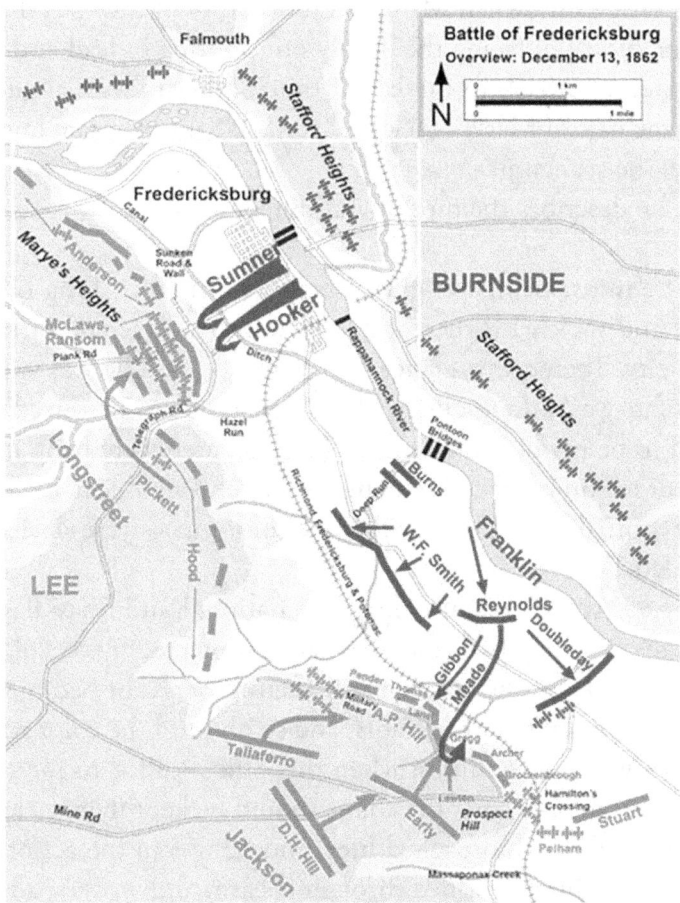

FREDERICKSBURG BATTLEFIELD

An Officers Observations

The planning was good. However, the timing and execution of the battle plan was sporadic and ineffective. And, as John, among others predicted, it was likely to fail. As Union General Ambrose Burnside began his assaults on the Confederates waiting along Marye's Heights, few could anticipate the scale of the slaughter this day would bring. Wave after wave of blue lines would surge into the maelstrom, only to be violently thrown back. Fifteen times, it is believed, Federal forces would march to that destination.

Among the fighting soldiers Colonel Frederick Hitchcock of the 132nd Pennsylvania led his men in their determined, yet vain, quest to reach the wall. Later when writing of the history of the 132nd he would discuss the ordeal of facing the deadly out pouring of lead and iron that awaited their approach to the lair of the Southern forces aligned confidently at the base of Marye's Heights behind what became an impregnable barricade, the stonewall.

Here he describes the most awful outcome of that fleeting, doomed attack:

"In the midst of that frightful carnage a man rushing by grasped my hand and spoke. I turned and looked into the face of a friend from a distant city. There was a glance of recognition and he was swept away. What his fate was, I do not know. That same moment I received what was supposed to be my death wound. Whilst the men were lying down, my duties kept me on my feet. Lieutenant Charles McDougal, commanding the color company, called to me that the color-guard were all either killed or wounded."

"We had two stands of colors, the national and State flags. These colors were carried by two color-sergeants, protected by six color-corporals, which made up the color-guard. If either sergeant became disabled the nearest corporal took the colors, and so on until the color guard were down. This was the condition when this officer called to me to replace these disabled men, so that the colors should be kept flying. He had one flag in his hand as I approached him, and he was in the act of handing it to me when a bullet crashed through his arm and wrist, spattering my face with his warm blood. I seized the staff as it fell from his shattered arm. The next instant a bullet cut the staff away just below my hand. An instant later I was struck on the head by the fragment of a shell and fell unconscious with the colors in my hand."

"How long I remained unconscious I do not know, possibly twenty minutes or more. What were my sensations when hit? I felt a terrific blow, but without pain, and the thought flashed through my mind, '*This is the end*,' and then everything was black. I do not remember falling."

"It takes time to write this, but events moved then with startling rapidity. From the time we went forward from the embankment until the line was swept back could have been but a few minutes, otherwise all must have been killed."

"When I revived I was alone with the dead and wounded. The line of battle had been swept away. The field about me was literally covered with the blue uniforms of our dead and wounded men. The firing had very perceptibly decreased. I had worn into the battle my overcoat, with my sword buckled on the outside. I had been hit on the left side of my head, and that side of my body was covered with blood down to my feet, which was still flowing. My first thought was as to my condition, whether mortally wounded or not. I was perceptibly weakened from loss of blood, but lying there I could not tell how much strength I had left. I did not dare move, for that would make me a target for the guns that covered that terrible wall, the muzzles of which I could plainly see. Many of them were still spitting out their fire with a venom that made my position exceedingly uncomfortable."

"What should I do? What could I do? To remain there was either to bleed to death or be taken prisoner and sent to Libby, which I felt would mean for me a sure lingering death. To make a move to get off the field would draw the fire of those guns, which would surely finish me. These were the alternatives. I carefully stretched my legs to test my strength, and I made up my mind I had enough left to carry me off the field, and I resolved to take my chances in the effort. I determined that I would zigzag my course to the rear so as not to give them a line shot at me."

"So getting myself together I made a supreme effort and sprang up and off in jumps, first to the right, then to the left. As I expected, they opened on me, and the bullets flew thick and fast about me. The first turn

I got a bullet through my right leg just above the ankle. It felt like the stinging cut of a whip and rather accelerated my speed. About fifty yards back was an old slab fence to my right, and I plunged headlong behind that, hoping to find shelter from those bullets. I fell directly behind several other wounded men, two of whom rolled over dead from bullets that came through the slabs and which were probably aimed at me. This flushed me again, and by the same zigzag tactics I succeeded in getting back to the railroad embankment, where, to my great joy, I found Colonel Albright with what remained of the regiment.

"Colonel Albright grasped me in his arms as I came over, with the exclamation, 'We thought you were killed.' Sergeant-Major Clapp told me that he had rolled me over and satisfied himself that I was dead before they went back."

"As I reached cover under this embankment I remember noticing a field-officer rallying his men very near us on our right, and that instant his head was literally carried away by a shell. So intense was the situation that even this tragic death received only a passing thought."

"Then came the Irish brigade, charging over our line as they did at Antietam. They came up and went forward in fine form, but they got but a few yards beyond the embankment, when they broke and came back, what was left of them, in great confusion. No troops could stand that fire. Our division and the whole Second Corps, in fact, were now completely disorganized, and the men were making their way back to the city and the cover of the river-bank as best they could, whilst the splendid old Ninth Corps was advancing to take its place."

Colonel Frederick L. Hitchcock, 132nd Pennsylvania Infantry

"War From the Inside"
J. B. Lippincott Company, 1904

Clara Barton at Fredericksburg

As the armies gathered, participant observers such as Clara Barton prepared to serve the fallen in their plight. She, like others, recognized the enormity of the venture and prophesied the terrible outcome likely to ensue. While she recognized the inevitability of a massive fight, like others, she also envisioned its cost in human souls. The evening before the terrible Battle of Fredericksburg, Clara Barton, wrote her cousin Vira a letter foreshadowing her predictions of the outcome on the following day.

Head Quarters 2nd Div.
9th Army Corps-Army of the Potomac
Camp near Falmouth, VA.
December 12th, 1862 - 2 o'clock A.M.

My Dear Cousin Vira:

Five minutes time with you; and God only knows what those five minutes might be worth to the many-doomed thousands sleeping around me. It is the night before a battle. The enemy, Fredericksburg, and its mighty entrenchments lie before us, the river between - at tomorrow's dawn our troops will assay to cross, and the guns of the enemy will sweep those frail bridges at every breath. The moon is shining through the soft haze with a brightness almost prophetic. For the last half hour I have stood alone in the awful stillness of its glimmering light gazing upon the strange sad scene around me striving to say, "Thy will Oh God be done." The camp fires blaze with unwanted brightness, the sentry's tread is still but quick - the acres of little shelter tents are dark and still as death, no wonder for us as I gazed sorrowfully upon them.

I thought I could almost hear the slow flap of the grim messenger's wings, as one by one he sought and selected his victims for the morning. Sleep weary one, sleep and rest for tomorrow toil. Oh! Sleep and visit in dreams once more the loved ones nestling at home. They may yet live to dream of you, cold lifeless and bloody, but this dream soldier is thy last, paint it brightly, dream it well.

Oh northern mothers wives and sisters, all unconscious of the hour, would to Heaven that I could bear for you the concentrated woe which is so soon to follow, would that Christ would teach my soul a prayer that would plead to the Father for grace sufficient for you, God pity and strengthen you every one.

"Mine are not the only waking hours, the light yet burns brightly in our kind hearted General's tent where he pens what may be a last farewell to his wife and children and thinks sadly of his fated men. Already the roll of the moving artillery is sounded in my ears. The battle draws near and I must catch one hour's sleep for tomorrow's labor. Good night dear cousin and Heaven grant you strength for your more peaceful and less terrible, but not less weary days than mine.

Yours in love,

Clara

True to her predictions Clara Barton served as a provider and caregiver to the men who fought at Fredericksburg on that fateful day. Moreover, she shared her observations with those prepared to listen, as well as those prepared to act on behalf of the men. That time quickly approached.

In December 1862, Clara Barton cared for the wounded from theBattle of Fredericksburg at the Lacy House (also known as Chatham Manor, previously an old slave plantation). As earlier Ms. Barton served in a variety of roles. During the Battle of Antietam, and later at Fredericksburg, she brought supplies to the front. At the later place she was assigned a room in the house that would enable her to offer her services in situ. There on December 11 she watched the bombardment of the town from the second floor. Shortly thereafter, as wounded men were brought into the house, she comforted soldiers from both sides and recorded some of her experiences in her diary.

She spent most of the following day at the Chatham Manor-Lacy House a place that became a hospital for the Union II Corps. The conditions were grueling. Since the doctors were too busy to keep medical records during battle, she recorded in her diary the names of the men who died at Chatham and where they were buried. The heaviest fighting of the battle occurred on December 13, and she spent most of that day in Fredericksburg attending to her charges, wounded men that surrounded her in the thousands.

Returning to Chatham, she spent the next two weeks at the mansion where, as she observed, the wounded occupied every room of the house and wrote that they "covered every foot of the floors and porticos." Adding to her notes, she wrote of soldiers that were sprawled out on the shelves of a cupboard, and the stair landings, commenting that a man "thought himself rich" if he laid under a table where he would not be stepped upon.

The task of caring for so large a contingent of wounded soldiers was overwhelming. So crowded was the existing facility that the 12,000 square-foot building did not contain enough space to hold all the wounded of the II Corps. Consequently, many men were placed on blankets in the muddy yard, where Barton set up a soup kitchen in a tent to help the wounded soldiers, as they shivered in the cold December air, waiting for someone inside to die and make room for them. It was truly a scene out of hell, a horrifying experience for patient and caregiver, alike.

Ms. Barton wrote of her many experiences at Fredericksburg in her diary, offering like other caregivers, special attention to the misconduct of officers whose responsibility for caring for their men was neglectful and shameless. In one sample she wrote:

"…But you may never have known how many hundredfold of these ills were augmented by the conduct of improper, heartless, unfaithful officers in the immediate command of the city and upon whose actions and indecisions depended entirely the care, food, shelter, comfort, and lives of that whole city of wounded men. One of the highest officers there has since been convicted a traitor. And another, a little dapper captain quartered with the owners of one of the finest mansions in the town, boasted that he had changed his opinion since entering the city the day before; that it was in fact a pretty hard thing for refined people like the people of Fredericksburg to be compelled to open their homes and admit these dirty, lousy, common soldiers," and that he was not going to compel it. This I heard him say, and waited until I saw him make his words good, till I saw, crowded into one old sunken hotel, lying helpless upon its bare, wet, bloody floors, five hundred fainting men hold up their cold, bloodless, dingy hands, as I passed, and beg me in Heaven's name for a cracker to keep them from starving (and I had none); or to give them a cup that they might have something to drink water from, if they could get it (and I had no cup and could get none); till I saw two hundred six-mule army wagons in a line, ranged down the street to headquarters, and reaching so far out on the Wilderness road that I never found the end of it; every wagon crowded with wounded men, stopped, standing in the rain and mud, wrenched back and forth by the restless, hungry animals all night from four o'clock in the afternoon till eight next morning and how much longer I, know not. The dark spot in the mud under many a wagon, told only too plainly where some poor fellow's life had dripped out in those dreadful hours."

From the *Diary of Clara Barton*

Letterman's Innovations

Somewhat earlier, after Antietam, but before the current battle, on October 30, 1862 Letterman established field hospitals while the Army of the Potomac was still in Maryland. It was a new system that would integrate the Ambulance Corps and medical supply support for upcoming battles. And it was soon tested at Fredericksburg, which offered these new arrangements a first opportunity to prove their worth. Corresponding with precautions implicit in John's warning regarding the forthcoming battle, Letterman was well at work implementing his new organizational ideas as the Armies came to do battle. And, as before, they proved successful.

A Surgeon's Observations

Among the observers of the scene at Fredericksburg, Surgeon Charles O'Leary, then Medical Director of the Sixth Corps, later President of the State Medical Society of Rhode Island, wrote in his official report:

"Being appointed Medical Director of the Sixth Corps a few days prior to the battle of Fredericksburg, December 13, 1862, I had the opportunity of putting in operation the Field-Hospital organization devised by the Medical Director of the Army, and witnessing its beneficial results. Within a very few hours after the positions were designated for the Field Hospitals on December 12th, all the necessary appliances were on hand, and the arrangements necessary for the proper care of the wounded were as thorough and complete as I have ever seen in a civil hospital. "During the engagements of the 13th, the ambulances being guided and governed with perfect control and with a precision rare even in military organizations, the wounded were brought without any delay or confusion to the hospitals of their respective divisions. Not a single item provided for the organization of the Field Hospitals suffered the slightest derangement, and the celerity with which the wounded were treated, and the system pervading the whole Medical Department, from the stations in the field selected by the assistant-surgeons with the regiments, to the wards where the wounded were transferred from the hands of the surgeons to be attended by the nurses, afforded the most pleasing contrast to what we had hitherto seen during the war.

"Both military commanders and medical officers agree that it would have been impossible for wounded to have received better care and treatment than they did in that battle." A similar state of things characterized the operations of the Medical Department in the rest of the Army."

11

THE WHISKEY SCANDALS

PAY DAY IN THE ARMY OF THE POTOMAC

"War is the time to profit from the mischief of man. It is the only good that can emerge from the carnage and suffering. So gentlemen, raise your glasses to salute this most welcome of opportunities."

Testament to the value of whiskey in the commerce of war:

> "Sir, you have asked my stand on the subject of whiskey. Well, if by whiskey you mean that degradation of the noble barley, that burning fluid which sears the throats of the innocent, that vile liquid that sets men to fighting in low saloons, from whence they go forth to beat their wives and children, that liquor the Devil spawns which reddens the eye, coarsens the features, and ages the body beyond its years, then I am against it with all my soul. But, sir, if by whiskey you mean that diadem of the distiller's art, that nimble golden ambrosia which loosens the tongue of the shy, gladdens the heart of the lonely, comforts the afflicted, rescues the snake-bitten, warms the frozen and brings the joys of conviviality to men during their hard-earned moments of relaxation, then I am four-square in favor of whiskey. From these opinions I shall not waver."

Anonymous Customer

The military commanders were, in large measure, disturbed by the abuse of alcohol among their men and expressed their concern in many ways. In his General Orders of February 4, 1862, General George McClellan admonished his troops that "total abstinence from intoxicating liquors…would be worth fifty thousand men to the armies of the United States." Recent court martial proceedings had brought the unwelcome admixture of soldiering and drinking to the general's attention and he hoped to alert his men "to the evils of intemperance and the terrible consequences of it to the individual soldier, no less than to the service." McClellan demanded, moreover, that all "Divisions, instead of imagining that these reflections are addressed to only one body of troop, will, with candor and magnanimity, accept their proper share of them, and resolve to contribute their exertions to the cure of this giant evil."

Opportunities for profiting from the war were numerous. Among them was the manufacture and sale of alcoholic beverages, where scarcity of product combined with an accelerated demand fueled an "ideal" climate for the operation of supply and demand economics among entrepreneurial men with little moral compunction or inhibitions, coupled with a thirst for the excessive profits to be gleaned from their customer's appetite for their goods. And of the latter, commensurate with their expectations, there were many as they soon discovered.

The Civil War was a factionist affair that tore the whiskey-making states apart. Pennsylvania a large producer of the product was solidly in the Union. However, elsewhere, distillers in places like Kentucky and Maryland, where alcoholic beverages were also manufactured, political attitudes often differed. These two States were two of the four border states in which slavery existed and maintained its legal status.

Since the late 1700s, when whiskey was first shipped down South, a number of Kentucky's whiskey-makers had come to rely heavily on the southern states' demand for whiskey to produce and market their products. Hence, people such as John Thompson Street Brown, father of George Garvin Brown (Old Forester), and the Weller brothers (W.L. Weller's sons), along with many other Kentuckians, served in the Confederate Army. Whether politically motivated, or profit-oriented, perhaps a bit of both, many such divisions occurred.

The production of whiskey was, like other enterprises, interrupted by the war. During the initial phases of the war some distilleries were destroyed, while some distillers died, and the remainder survived as best as they could. However, whatever the individual circumstances the war changed the normal course of enterprise fostering scarcity of the product, and most importantly, enhancing its value as a source for raising revenue. As early as 1862, President Lincoln was forced to reintroduce the excise tax on whiskey (abandoned earlier) to help pay for the Union war effort. Once again, just as in the Revolutionary War and the War of 1812, the production of whiskey was encouraged to help finance the armed forces.

Among the varied enterprises (supplying uniforms, shoes, weapons, food and transportation, among other products and services) that engaged entrepreneurs, whiskey had substantial value during the Civil War. It had the power to soothe men's souls, to make them forget the carnage that they encountered on the battlefield, and perhaps most importantly, whiskey often acted as the only anesthetic available to help soothe and heal the wounds caused by war. This was especially true in the South.

During the Civil War, whiskey was widely used as an antiseptic for wounds sustained on the battlefield, as well as poured down parched throats to suppress both awareness and ease the pain of countrymen fighting countrymen on their own soil.

Northern soldiers had more money than their Southern adversaries and could buy more whiskey when it was available. But although Union officers were allowed to buy whiskey, enlisted men had to rely on rations that were sanctioned by their commanders as their legitimate source of liquor. Needless to say soldiers on both sides were, for the most part, hungry, cold, frightened and sorely in need of solace wherever they could find it. If temporary refuge from their plight lay in a slug of whiskey, they would find a way of obtaining it.

Union troops procured their whiskey from wherever they could, by legal and illegal means: having it sent by their families, dodging the guards and finding their way to a local grogshop and, in the case of one enterprise involving the whole regiment during the Christmas celebrations of 1864, making a full 15 gallons of whiskey all by themselves. Confederate troops, on the other hand, did not get their fair share of whiskey, not only because of their lack of hard cash but also because the South couldn't afford to use what valuable grain there was to make whiskey; people were often lacking the basic necessities for life (such as grain) and the scarcities of food products demanded by the war left them and their animals in poor straits.

Whatever the reality of assertions concerning who was drinking more, (and there were claims on both sides) the Southern people needed food more than they needed whiskey. The Confederacy, therefore, declared prohibition on a state-by-state basis and tried to buy up all the available whiskey it could to use as medicine. As would be expected, the States reacted to the prohibition of whiskey with varying degrees of complicity. The fact that whiskey was declared illegal and was hard to come by created changes in the marketplace. Most significantly, its value increased many fold over time. As the War continued scarcity increased and the black-market price for whiskey rose substantially; in 1863 it was priced around $35 per gallon, while three years earlier prewar in1860 the cost of a gallon of whiskey was about 25 cents for the same quantity of product.

Black marketers who had the means to make whiskey found a way to exploit the trade to considerable advantage and did so. Overall, the Civil War's effect on the whiskey business, by no means negligible, was to whittle down the number of whiskey distilleries and distillers, a fact that probably didn't upset temperance advocate Abraham Lincoln. But in the process it also spurned a "new" industry promising significant profits for its investors and those willing to accept the risks associated with its production and distribution. Most significantly, it led to many abuses, the most common being alcoholism and alcohol-induced behavior characterized by drunkenness, dereliction of duty, the mistreatment of prisoners, and occasionally, to theft and murder. Furthermore, the rush to secure profits led to inferior and poorer products being produced. And this, in turn, resulted in serious health consequences among the many men that consumed the product. The problem was exacerbated over time.

A Meeting of Dr. Letterman and John Scobell

Shortly after the Battle of Fredericksburg Dr. Letterman, on the suggestion of Major General Hancock approached John Scobell to discuss a means of curtailing the rampant distribution of harmful whiskey to the troops.

In their last meeting Hancock was dismayed. He was distressed over the increasing incidence of alcohol related-incidents involving the troops. In his response to a recent memo from Letterman he noted, "As your last report indicates, the problem of the men's drinking has gotten out of hand. As you have observed, they are frequently late in reporting to duty, neglectful in the completion of their tasks, and border on the insubordinate in relation to their superiors. This problem has been compromised and compounded lately by the poor quality of spirits that they have been consuming causing medically related problems. Some, in increasing numbers, as you know, have shown signs of alcohol poisoning and addiction."

At this point Scobell indicates, "These observations are correct. I have noted a number of incidents in both the white and black communities that support your concerns."

"Well", adds Letterman "the general has suggested that we could use your help in addressing this problem. He proposed that you go underground among members of the black community to see if they can offer information that we may employ in stemming its influx."

In response, Scobell asks, "How can that help?"

To which Letterman answers, "The general has offered the following comments. It appears to him that white men that previously may have been employed in the production of whiskey under normal circumstances are now less available for its production. He notes further that the work of distilling spirits is arduous and suspects that black men, both free and slave, are being used in its manufacture at this time. Our history of misusing black people in performing undesirable work such as this would suggest that they would be involved on the most basic level. If we can learn from them the names of the persons responsible for this mess, that is those involved in its funding and production, there might be grounds for restricting its practice."

Somewhat puzzled John asks, "I don't understand. How can that information assist us? Even if we can extract this information from the producers?"

"It is a bit complicated, but not irrational", responds Letterman. "There are taxes that have been imposed on the manufacture and selling of spirits to help pay for the war effort. At this point in the war the government depends on these revenues to fund our military activities. Certainly, a case of tax fraud can be made against illegal producers and distributors of alcohol many of whom have been negligent in paying their fair share of taxes. That could put them out of business."

"And, as you know", he continues, "There remain a number of unsolved and suspicious murders of soldiers involved in the trade. Perhaps, if we are careful we can make cases against these persons for these horrible crimes, as well."

Somewhat enlightened by Letterman's explanation John says. "You make an appealing case, doctor. And as you indicate, the crimes being committed affect both the white and black communities. Perhaps, some discrete inquiries by Jed and myself are in order". And as he and the doctor part, agreeing to meet within a fortnight, John suspects that his latter words will prove prophetic.

12

Perpetrators of the Whiskey Schemes

The Schemes of Leroy "Bull" Brown and Leander Worthington

"All that I know is that when men gather, they drink. And they drink a lot. And there is profit to be made from drink."

Leroy "Bull" Brown

They were an unlikely pair; Leroy was a black man that achieved minor fame as a fighter. At 6-3 in height weighing 240 pounds he was certainly bigger and stronger than your average man. He was also meaner than most. Leander Worthington, on the other hand, was neither bigger nor meaner than most men. Rather, he was a society figure, who had accumulated considerable wealth and power by virtue of his family's political connections. But this unlikely pair had something in common, an enterprising spirit, especially in the production and distribution of alcohol to members of their own race. Profit did not distinguish between men based upon the color of their skin. And both men were bend on extracting profit from the weaknesses of their clients.

Early in the game when they met, Bull was on the production end of the scheme and Leander on its distribution. Moreover, they worked different markets, both black and white. However, they soon learned that there were few differences among the markets that they served; demand for their product was comparable. And each of these men was uniquely suited, by race and temperament, to cater to that market most aligned with the black and white consumer of their respective races.

Bull, through his control over black workers soon found them to be willing consumers of the product that they produced, one step away from its distribution and consumption. Worthington, with Army connections secured early at the outbreak of the war, became a major provider of drink to camp sutlers who plied their trade, including whiskey, among the soldiers.

The scarcity of food for both Confederate soldiers and Confederate civilians was directly related to several factors, factors which worsened as the war progressed. Among contributing factors, Union blockades kept both grocery items and seeds from reaching the Southern states. As supplies of both groceries and the means for increasing stores diminished, the prices of these items soared. Inflation, however, was only part of the problem; by the last years of the Confederacy, Confederate-issued currency was worth less than the paper it was printed on, and many of the merchants who did somehow acquire groceries, seed, and other food would not accept the currency of their own nation, exacerbating the problem. Bartering of products, such as whiskey and coffee, subsequently came into vogue.

The lack of reliable transportation in the South also contributed to the difficulties in the distribution and sale of goods. Railroads, rife in the North, were just making inroads in the South at the outset of the war, with most work ceasing when the fighting began. With no reliable long-range transportation system, fresh food and other food items could not be feasibly transported to either soldiers or civilians throughout the South. During these tumultuous times whiskey, in addition to other products became a scarce commodity and was prized. In addition to other goods it was subject to exploitation.

Profiteering in the North

As John met with Jed, shortly after his encounter with Dr. Letterman regarding the problem of alcohol consumption he outlined the task set before them by Dr. Letterman.

The business schemes at the time were often outrageous. As they knew by rumor and innuendo, and in spite of the lofty goals espoused by supporters of the war, John and Jed found that there were many profit-hungry entrepreneurs seizing upon opportunities offered by the war. Moreover, they were witness to a variety of the corruption schemes that characterized war profiteers. From inferior uniforms to ill-fitting shoes to inedible food and drink the diversity of schemes proposed through ill-gotten profits was astounding. Many were easily discovered over time, while others proved less evident and more intractable.

Included among those who plied the military trade as profiteers were Elisha Brooks and his brothers representing the firm that achieved prominence as clothiers bearing the label of Brooks Brothers. The brothers were the first recipients of a Government military uniform contract, an invitation with negligible oversight that provided them opportunity for mischief. Their firm, in response to the demand for military clothing produced uniforms that were turned out in a matter of weeks and issued to the new recruits. At face value their business appeared legitimate. However, there were serious questions surrounding the quality of their products; jackets, for example, were missing buttons and featured button holes that were barely threaded together. Moreover, the uniforms fit poorly and were made from a wool substitute consisting of sawdust, scraps and threads, glued and ironed together, often falling apart in the first rain. A total of 48,000 of these uniform frauds were paid for, before a government board of inquiry required Elisha Brooks to testify why he had used sub standard materials. To these charges, Brooks offered a trivial explanation to his inquirers, to which he personally assumed no responsibility.

In the meantime Brooks Brother's had contracted to supply other military items including, belts, shoes, tents, felt hats, ponchos and blankets.

A reporter for the *New York Tribune* termed the accouterments, "Shoddy, poor sleazy stuff, woven open enough for sieves, and then filled with shearman's dust… Soldiers, on the first day's march or in the earliest storm, found their clothes, overcoats, and blanket, scattering to the wind in rags or dissolving into their primitive elements of dust under the pelting rain."

Brooks Brothers manufacturing and warehouses did not escape public criticism, nor retribution as punishment. In partial response to their practices their factory was looted and burned during the draft riots of 1863 in New York City, a token payment one might conclude for their "shoddy" treatment of the men that were bleeding and dying in order that Elisha Brooks and his brothers could be guaranteed a profit.

Ironically, the strength and persistence of the Brooks Brothers label survived the war, as did many scams of the period. Among the many ironies of history it may be noted that President Abraham Lincoln wore a Brooks Brother's suit when John Wilkes Booth assassinated him. Moreover, over time Brooks Brothers suits gained fame in spite of their questionable origins. Today the brand is internationally known and its products are sold worldwide, often worn by celebrities and Presidents, alike.

Profiteering in the South

Concerns about war profiteering were not limited to the activities of a few "shoddy" millionaires in the North. In the Confederacy, where supplies were severely limited, and hardships common, the mere suggestion of profiteering was considered a scurrilous charge. Georgia Quartermaster General Ira Roe Foster attempted to stimulate the supply of material to Confederate troops by urging the young women of his state to knit 50,000 pairs of socks. Foster's sock campaign served as an incentive increasing the supply of this needed item, but it also met with a suspicion of "wrong-doing" and a corresponding backlash. Whether a suspected Union disinformation campaign launched at the time, or the work of suspicious minds, rumors abounded of his malfeasance (neither beyond allegation nor definitively proven), and which Foster denied as a "malicious falsehood!" spread that he and others were profiteering from the sale of the socks. It was alleged that contributed socks were being sold, rather than given freely to the troops. In response, Foster undertook a newspaper campaign to attack the unfounded rumors, and to encourage the continued contribution of socks for the troops. He offered $1,000 to any "citizen or soldier who will come forward and prove that he ever bought a sock from this Department that was either knit by the ladies or purchased for issue to said troops."

The Whiskey Conspiracy - Setting the Trap

The manufacture of bootleg whiskey grew in proportion to the expansion of the war effort. Over time, as it was brought to their attention, the problem was now one in which Dr. Letterman and General Hancock realized could no longer be ignored. Subsequently, in response to the growing epidemic of misuse of alcohol and a proportional growth of alcoholism, John and Jed were assigned by Major General Winfield Scott Hancock to investigate the issue and find a way of limiting the activities of those involved in the scheme.

It is now somewhat later where we find John and Jed discussing the issue, hoping to form and construct a plan of action. At this point they have agreed that their plan should be simple and lead to the discovery of facts that are not only authentic, but could also stand the test of scrutiny in a white man's court; as they knew such courts excluded the testimony of black men limiting their power to evoke judicial process. They also knew that the probability of completely disrupting the whiskey trade was slim: it was far too vast an enterprise to be eliminated entirely. Moreover, the likelihood of them holding the top man responsible, when identified, would also be improbable; he would have to convict himself.

Shortly after their discussion, two inebriated soldiers returned to camp after a short furlough. They are staggering and disoriented, barely recalling the two-day bender that they had just embarked upon. They are also suffering from a variety of alcohol-induced symptoms suggesting long-range consequences stemming from their consumptive habits. It is a familiar pattern, frequently observed earlier.

In response to their plight, the attending regimental physician called upon to treat them brings their case to the attention of Dr. Letterman. Distressed by the results stemming from his examination, Dr. Letterman seeks out John Scobell to deliberate upon the situation with him.

"Mr. Scobell, the case of the two soldiers that I have recently attended suggests that the problem of alcohol poisoning is more serious than originally suggested. We need to find a solution to this problem and to do so fast."

"Yes, Major, I hear the urgency in your voice and I readily agree. Do you have any information that offers us the source of their supply of alcohol? Jed and I have made some inquiries and we have located a Mr. "Bull" Brown a black man who is a major supplier to sutlers that sell prohibitive beverages to enlisted men. However, Bull Brown is not, as far as we can determine, the primary source of this illegal alcohol."

"Yes, there are others over him. One name that has been suggested is Leander Worthington. What do you know of him?"

"Well, he appears to be a major conduit for illegal whiskey in these parts. But, unfortunately, we have not been able to connect him with these operations. From what we have been able to determine is that he operates with impunity. According to rumor Worthington is one of the top distributors of alcohol to men like Brown where the latter has been operating in the black community for many years. Apparently, he is protected by some higher powers possibly a key supplier such as Worthington. Brown has now expanded into the white community. If we can only find proof that Worthington is one of his protectors, as well as his supplier, we might be able to turn off this particular faucet of illegal whiskey at its very source."

"For now, we may be able to have the provost general's office arrest Brown for violation of army provisions regarding the distribution and sale of whiskey to enlisted men. When we have him in custody we may be able to sweat him to reveal other names associated with his activities."

"It sounds like a good plan. I will arrange to discuss it with General Hancock. If he agrees then we can proceed"

Two days later Dr. Letterman meets with John and Jed once more. "Gentlemen, General Hancock has agreed to your plan. We have his permission to arrest Brown and place him in Army custody."

The next day John and Jed spend the morning in the black community, learning among other things, the whereabouts of Bull Brown. Later that day John and Jed, fully armed confront Bull Brown in an alley outside his home. As Brown is returning from his business warehouse, about a half-mile from his residence he is surprised to find John and Jed, with weapons drawn ordering him to stop.

Realizing that the two men confronting him are armed, and that he is at a disadvantage he offers them no resistance. Instead, he says, "I have no quarrel with you boys. If you will release me, perhaps we can come to some accommodation. I'm sure that you can profit from this miscarriage of justice."

Recognizing this attempt at bribery for what it is John says, "Not this time Mr. Brown. We have a warrant for your apprehension and arrest. You will be spending the night in the stockade and hopefully, significant time afterward."

One Week Later. A speedy trail is requested and granted. Within a fortnight Bull Brown is placed before a military tribunal. While Brown's attorney requested a civil trial, where his influence could be pressed, his client is denied a trial in civilian court; alternatively, he was knowingly placed under the scrutiny of a military tribunal less subject to public influence.

In addition to Brown, a variety of people, several suspected as being involved as illegal producers and distributors of whiskey, are ordered to appear in court. Among them, Leander Worthington is ordered to present himself before the tribunal in order to give testimony.

The Trial

On the day of the trial the courtroom is filled with people, among them journalists anticipating a big story. As we join the proceedings, Bull Brown is in the processes of reluctantly giving testimony. He is facing the military adjutant and has just been asked whether the set of books seized at his home and later carefully examined by the prosecution represents the accumulation of his accounts.

Bull squirms in his chair. He recognizes that he is in a precarious position. Under the present circumstances he is likely to be convicted of the crimes he has been charged with committing. His only hope now is to reveal the names of his confederates, and hopefully, pass on to them some of the blame, lessoning his own sentence. The prosecutor has assured him that by turning states evidence he would facilitate his wish to plea bargain. It was a chance he had to take.

Under the judges prompting the prosecutor asks him again, "Mr. Brown is this journal that I hold in my hand the only ledger of your accounts in the whiskey trade?"

In response, Bull testifies, "Yes but that is not the only set of books. There is another with more detailed information, including the names of others involved in this project."

At that moment one of the spectators seating in the first row quickly rises from his seat and shouts, "You bastard! You have been keeping two sets of books. Why you cheating black bastard. You have been stealing from me" And before his lawyer can restrain him, Leander Worthington inadvertently reveals his complicity in the trade as he shouts out "I trusted you. We were partners and you violated my trust".

Upon this revelation there is uproar in the courtroom and the presiding judge starts to bang his gravel on his desk in order to restore decorum within the courtroom. He then yells to the bailiff, and says, "Take that man out of this room and place him in custody."

13

THE DEVELOPMENT OF NEW WEAPONS

The Civil War was a time of great social and political upheaval. It was also a time of great technological change. The efforts of inventors and military engineers led to the invention of many new kinds of weapons. These included the repeating rifle and the submarine, weapons that in this war and those that occurred later changed forever the way that wars would be fought. Equally important were the technologies that did not specifically have to do with the progression of war, such as the development of railroads and the telegraph, public inventions, that while not a response to war, advanced its cause. Innovations like these did not just change the way people fought wars–they also changed the way people lived.

New Kinds of Weapons

Before the Civil War, infantry soldiers typically carried muskets that fired one bullet at a time. The range of these muskets was about 250 yards. However, a soldier trying to aim and shoot with any accuracy would have to stand much closer to his target, since the weapon's "effective range" was only about 80 yards. Therefore, armies typically fought battles at a relatively close range. Rifles, by contrast, had a much greater range than muskets did; a rifle could shoot a bullet up to 1,000 yards and was significantly more accurate than its predecessor. However, until the 1850s it was nearly impossible to use these guns in battle because, since a rifle's bullet had roughly the same diameter as its barrel, they took too long to load. (Soldiers sometimes had to pound the bullet into the barrel with a mallet.) However, all that soon changed.

In 1848, a French army officer named Claude Minié invented a cone-shaped lead bullet with a diameter smaller than that of the rifle barrel. Soldiers now could load these "Minié balls" quickly, without the aid of ramrods or mallets. Rifles with Minié bullets were also more accurate, and therefore deadlier, than muskets were, which forced infantries to change the way they fought: Even troops who were far from the line of fire had to protect themselves by building elaborate trenches and other fortifications.

Rifles with Minié bullets were easy and quick to load, but soldiers still had to pause and reload after each shot. This was an inefficient and dangerous process, especially during battle. By 1863, however, there was another option: the so-called repeating rifles, or weapons that could fire more than one bullet before needing to be reloaded. The most famous of these guns, the Spencer carbine, could fire seven shots in 30 seconds.

Like many other Civil War technologies, these weapons were more readily available to Northern troops than their Southern counterpart: Southern factories had neither the equipment nor the knowledge of how to produce the advanced weapons being developed in the North. These disparities did not go unnoticed. As one soldier remarked: "I think the Johnnies [Confederate soldiers] are getting rattled; they are afraid of our repeating rifles." Adding to this observation, one Union soldier wrote. "They say we are not fair, that we have guns that we load up on Sunday and shoot all the rest of the week."

The Gatling Gun

In addition to many of these innovations, perhaps the most deadly of ground weapons to be developed was the Gatling Gun, an instrument, if properly deployed, that was capable of out performing most weapons within any existing arsenal of weapons at the time, producing mass destruction in its wake. It was truly a gun offered before its time, but not all time.

Ironically, according to its inventor, R. J. Gatling, the gun was developed as a deterrent to battle, in order to discourage war, making its destructive power a cause for pause, rather than a formidable weapon of choice. The power of the weapon was, according to its inventor directed toward furthering that goal as no other weapon at the time could. In a letter outlining his goals and hopes for the gun, Gatling wrote:

My Dear Friend.

It may be interesting to you to know how I came to invent the gun which bears my name; I will tell you: In 1861, during the opening events of the war, (residing at that time in Indianapolis, Md.) I witnessed almost daily the departure of troops to the front and the return of the wounded, sick, and dead. The most of the latter lost their lives, not in battle, but by sickness and exposure incident to the service. It occurred to me if I could invent a machine--a gun-- which could by its rapidity of fire, enable one man to do as much battle duty as a hundred, that it would, to a great extent, supersede the necessity of large armies, and consequently, exposure to battle and disease be greatly diminished. I thought over the subject and finally this idea took practical form in the invention of the Gatling gun."

Yours truly,
R. J. Gatling

The Gatling gun's operation centered on a cyclic multi-barrel design which facilitated cooling and synchronized the firing/reloading sequence. Each barrel fired a single shot when it reached a certain point in the cycle, after which it ejected the spent cartridge, loaded a new round, and in the process, allowed the barrel to cool somewhat. This configuration allowed higher rates of fire to be achieved without the barrel overheating.

The gun, as was common at the time, was named in honor of its inventor, Dr. Richard Jordan Gatling, a physician. Gatling was apolitical in his sympathies concerning the Civil War. Yet, while trying to sell machine guns to the Union, ironically, he was an active member of the Order of American Knights, a secret group of Confederate sympathizers and saboteurs.

Act 1 – A Proposal is Set Forth

The Order of American Knights

Many northern citizens opposed the Civil War. However, these opponents usually resorted to peaceful forms of protest to object to the war and referred to themselves as Peace Democrats. Their opponents, however, referred to them as Copperheads after the venomous snake they reputedly imitated.

There were numerous reasons why the Peace Democrats opposed the Union war effort. Many Copperheads were former residents of the South and had family members still living in the seceded states. They feared, as a result of the war, that Northern military efforts would harm their loved ones, as well as destroy or confiscate property among those who remained in the South. Other Peace Democrats feared that President Abraham Lincoln intended to free the slaves in the South. If Copperheads had slave-owning family members still in the South, their relatives stood to lose sizable amounts of wealth if the federal government intervened and ended slavery.

Many Peace Democrats without slave-owning relatives in the South also feared slavery's termination on other grounds. These opponents believed that with emancipation large numbers of freed slaves would relocate moving from the south to the north, causing massive job losses among whites and a decline in property values wherever the freed African Americans eventually settled. Other Peace Democrats objected to President Lincoln's suspension of habeas corpus, including the right to be charged with a crime and the right to a speedy trial, while those with economic ties to the South, feared that their businesses would suffer while the Civil War raged.

The Order of American Knights eventually formed in the North to protest the Union war effort. Most of this group's members were originally Peace Democrats, but as the Civil War continued and Southern victory became less likely, they adopted a more radical approach, including violence and sabotage, to protest the conflict. During the Civil War, Ohio authorities claimed that between eighty thousand and one hundred thousand Ohioans belonged to the Order of American Knights. However, more recent evidence suggests that significantly fewer Ohioans belonged to this group.

While most Northerners supported the Union war effort, Copperheads were a sizable and vocal minority. Their actions, especially those of the more extremist Order of American Knights, prompted a great deal of concern among Northern officials during the war's first years. By late 1864 however, Union military victories and the nearing end of the conflict caused most Northerners to rally behind President Lincoln and the Northern war effort. The more violent actions of radical Copperheads, including the Sons of Liberty, also caused Peace Democrats increasingly to unite in their sympathies with the federal government.

During the early stages of the War many meetings of members of the Order of American Knights took place. While these gatherings were initially designed to increase its membership, as the war continued these meetings took on an ominous tone. In one such gathering in the Midwest that is detailed below, a dramatic proposal was offered regarding the new Gatling gun. As the scene is assembled:

Among the 150 assembled delegates a proposal is introduced by a speaker, a gentleman from Ohio. "Mr. Chairman, I propose that we authorize the purchase of several new Gatling Guns to be manufactured by the Liverpool Iron Works. These guns could turn the tide in the favor of the South and bring the War to a righteous conclusion."

Another delegate, excited by the murmur of the crowd upon hearing this proposal, quickly rises from his seat and says, "But we do not have possession of the plans for this miraculous gun that I have heard tell. I would agree that at this stage of the war, a super weapon would be desirable. Yet, how can we obtain one or the plans to manufacture one? And who, or where, can we do this? At this point in the war how can the South's dysfunctional manufacturing section build an effective weapon of this sort? Or, is it likely that it could?

At this stage of the discussion there is further mutterings from the crowd. They have been stirred up for action. A young inventor in the assembly now rises to offer his opinion. "If plans could be secured", he says, "or a weapon seized to be used as a prototype then we can find a sympathetic manufacturer in Europe that should be able to manufacture a sufficient number of these weapons to aide our cause."

As the crowd signals its approval through murmurs and heads nod, the chair hammers his gavel and declares, "All those in favor of this proposal indicate your approval." And with that impetus as a backdrop, the crowd shouts its consent.

Act 2 - Support for the Plan

At the President's Mansion in Richmond.

It is after dinner at the Confederate President's manor in Richmond Virginia. Before the men adjoin to the smoking room, as is customary, Varina Davis asks the president and their remaining guest Judah Benjamin Secretary of War, their close friend and closest and most trusted presidential advisor, to remain after the others have retired.

The relationship between this trio of political allies and personal friends was intense and sincere. In addition to his special relationship with the President, Benjamin was very close to the Confederate First Lady

Varina Davis. Often, they would exchange confidences regarding war events and the latest information on the President's health. It was reported that, "Together, and by turns, they could help him over the most difficult days." This was one of those times.

"Mr. President," says Benjamin, "It has been brought to my attention that the North has purchased a so-called Gatling Gun and that this war instrument can enable one man to fight better than 100. Also, it has been proposed that we secure plans to build a number of these guns. As Ms. Davis and I have discussed, with our limited and diminishing supply of men, this instrument could alter the course of the War in our favor." Pausing momentarily, the Secretary continues, "If applied properly, of which I have no doubt regarding its effectiveness, it should be able to bring the North to the peace table and be willing to negotiate with us as a separate nation, a condition that President Lincoln refuses to accept."

"Yes, love", says Varina. "Mr. Benjamin and I have consulted on this matter. We think that we should pursue a course that will enable us to secure this weapon and then negotiate with the English for a number of them to be manufactured there and shipped to an operating port."

Surprised at the suggestion, but nevertheless receptive, the President turns to them and says, "But as you both know, we are nearly bankrupt now. How can we pay for these new weapons? Would our creditors be willing to extend to us additional credit for this purpose?"

"Well, Mr. President we can barter with them," suggests Benjamin. "And preliminary signs suggest that they are willing to listen to our proposals. My source in London indicates that they would be open to a long term arrangement that benefits the both of us."

"And what would be the nature of this arrangement, Mr. Benjamin?"

"As you know, the mills of Britain, developed during the first half of the 19th century, use more cotton than the rest of the industrialized world combined. At this point, cotton imports to Britain have come almost entirely from the American South.

"Aside from the rhetoric, we need to recognize that this is a vital commodity to them.

"According to an article printed in The Economist in 1853, one commentary said, "let any great social or physical convulsion visit the United

States, and England would feel the shock from Land's End to John O' Groat's. The lives of nearly two million of our countrymen ... hang upon a thread."

"It appears to Ms. Davis and myself that if we can guarantee to provide the English with a specific and exclusive quantity of the product for 10 years after the war has ended (a stipulation that would offer them an economic and political advantage over the French), that they would be willing to extend to us the necessary credit to pursue the war to a just end."

Upon hearing this, Davis becomes pensive. Then he says, "In some ways that would be an ideal arrangement. Aside from the immediate problem that we face, it would insure us a market after the War and later, as well."

"Let us say that the Treasury advances your department $50,000. You may use the money at your discretion to pursue a course of action where we can obtain a prototype of the gun. How will it be manufactured? ? I suspect that your contacts in England will be of assistance to us in this matter. I will leave the details of that task in your able hands, Mr. Benjamin."

At this time Benjamin had many associates in London. One of these was Henry Hotze a Southern propagandist journalist and publisher. Hotze, he decided, would introduce Benjamin's agents to influential British politicians and industrialists that would manufacture the Gatling guns.

HENRY HOTZE

From 1855 to 1865, Hotze was an associate editor for the Mobile Register. He also served in the Confederate Army and as a Confederate diplomat. Moreover, he edited a London-based newspaper that promoted the Confederate cause. His writings promoting a "scientific hierarchy of race-based intelligence" were used by white supremacists as a justification for slavery. A somewhat overlooked figure of the American Civil War era, Hotze was arguably the South's most effective, if not tireless propagandist abroad.

And so it was done. First, through diplomatic dispatch, then through direct contact with agents sent to London by Benjamin, contacts were made via the British government with ship builders that would be willing to copy a prototypical Gatling gun. Because they would be willing to undertake the project with short notice, these manufactures were able to negotiate excellent wages and hire the top craftsmen they could find. They would stand by to execute the contract when the time came.

As early as 1860, Thomas Jordan secretly initiated a pro-Southern spy network in Washington, D.C. Somewhat later, in early 1861, Jordan passed control of the espionage network to Rose O'Neal Greenhow; moreover, he continued to receive and evaluate her reports after she was restricted to house arrest in August 1861 and imprisoned in Washington, DC in January 1862. He appeared to be her Confederate Secret Service "handler" during this formative phase of Confederate intelligence.

Jordan would be an invaluable man to lead a small group of co-conspirators in obtaining a Gatling gun prototype. He was a West Point graduate with Army experience and multiple contacts.

Moreover, he was born in Luray, Virginia, home of the Luray Caverns, an ideal place to secretly store the guns manufactured in England until they were called upon for employment.

Jordan's plan was simple. He would direct a high jacking attempt and seize a gun that would be disassembled locally, and whose parts would then be sent to England for replication. His network of spies learned that Hancock had ordered a number of guns to be delivered to his Corp. In addition, he knew that they would be shipped to Virginia by train. His plan was to intercept the train carrying the gun, likely to be shipped as single units, one gun at a time, as a security precaution. But one gun was all he needed.

Act 3 –Preventative Intervention

Knowledge of the purchase and manufacture of the gun was especially tight. However, information regarding details of its shipment was not. And this worried Major General Hancock who expressed his concern to John Scobell. In one of their informal meetings he articulated this concern to John.

"Mr. Scobell, we have arranged to have several of these new repeating guns sent here. It is a revolutionary weapon that can have a major impact upon our hopes for changing the course of the war. However, I am concerned that information of its shipment has fallen into enemy hands. There are simply too many people involved in this process. And, if that is the case, as I suspect it is, then we have a problem. For security's sake we have taken several precautions. Each of the three guns we have ordered will be shipped separately; they will arrive by independent trains within a fortnight. Yet I am concerned. We will need to insure ourselves that the enemy does not intercept any of these shipments. We cannot station troops along the entire track for the distance will be too great, stretching our resources too thin. I believe that we need a plan. And I need your help in implementing it. How shall we proceed?"

"Well, Sir! I can suggest a radical proposal that might work. Perhaps, we can send a newly trained group of armed black troopers on every train to accompany each gun. I can supervise the first shipment in case that is their target, while Jed could supervise the second shipment. In the event that neither shipment is attacked, I could double back and oversee the third shipment. My suspicion, however, is that the enemy will act quickly and try to seize the first shipment if they are well organized. And if not, they will try to seize the second. As ourselves, they work under time constraints, favoring quick action."

"Mr. Scobell, I think that your reasoning is correct and we shall proceed according to the plan you outline. I will have my adjutant draw up orders to request a company of Colored Troops that you can split into three units for this assignment. He will also assign them to each train, as you dictate, to be on alert to surprise attack."

Within two days the plan is set and the requisite soldiers are assigned under John's direction. They are a mixed group of warriors, some with experience, others raw recruits.

It was an organizational plan routinely used by the South. While the North often replaced whole regiments of troops for assignment, the South employed a different approach; they would integrate new troops within the core of existing units, insuring uniformity of performance by allowing experienced troops to guide those new to the ranks. In the absence of time for proper training, a condition found throughout the war on both sides, and on many fields of battle, those with little experience would learn soldiering at the side of those seasoned troops among them.

John chose to ride in the first train carrying a Gatling Gun. He took 30 troops with him, evenly divided between those with battle experience and those without. Among the former was Sgt. William Marshall, a freed slave from Virginia. Up until a year ago William a tall brown skinned man with a soft manner belying his fierce heart, was a slave. More recently he was a soldier that had gained experience as the Army of the Potomac began to recruit black soldiers. Although born a slave, he fervently believed that this was his country and he hoped to liberate it from the hated institution of slavery. According to John's instructions, William selected 14 additional men for guard duty to help in deterring any attempt on the theft of the gun. In reviewing his decisions John thought that he selected his men well.

On the third day out of its originating depot the train entered Virginia. It had traveled some distance and it had not experienced any difficulties; there was still no sign of trouble. However, that was all about to change shortly. Midday into the third day, John and his colleagues heard a mighty rumble accompanied by an earth roar as the train slowed to round a prominent ridge. It was reverberations of a canon firing shells in the vicinity of the train; at that point they were under attack.

Within a few minutes a shell landed in front of the engine, tearing up a section of track. The damage caused was sufficient to stop the train and only an alert engineer was able to avoid catastrophe. At that point five riders each on horseback, followed by an accompanying wagon moving quickly toward the stranded train entered the scene. They were in pursuit of the freight car prominently carrying the gun peering from its roof. This was the attack that John had anticipated.

With armed men lined up on both sides of the doors to the freight car, John issued an order, transmitted by Sgt. Marshall to his men, to open fire as the assailants pried the doors open. From that point on mayhem ensued and the exchange of shells that followed was rapid and deadly. Four of the assailants were hit by oncoming fire and thrown from their saddles, while the fifth, sensing the unexpected, swiftly turned his horse and sped off in the direction from which he came. In the exchange one trooper was killed while Sgt. Marshall was wounded. However, the gun was saved.

In the aftermath of the fighting William is taken to the nearest federal hospital. Unbeknownst to him initially, his fiancé, Louisa Curtis rushes to be near his side and to serve as his nurse. Several weeks after his condition has stabilized and he is no longer in danger, she and William engage in the following conversation as Louisa says, "I have been informed that when you are fully healed you may return home and I hope that you will come home then. Perhaps, we can then go about resuming a normal life".

However, William is abrupt on his response.

"No, I cannot," says William, "I must remain with my men. This fight is far from over and I need to attend to its conclusion."

Among the visitors of the sick and injured, John approaches William's bed, and overhears the conversation between William and Louisa. John recognizes that he has interrupted a private, unshared conversation between these two lovers. Normally, reticent in such matters he, nevertheless, is inclined to speak on this occasion. He leans over the bed where William is resting and says to him, "No, William. You have served your country. And you have done well. This war is not over and likely won't be for some time. However, now it is time for you to go home, to rest and recuperate, as I hope we all may do at sometime in the near future."

14

ON TO CHANCELLORSVILLE

First Confederate soldier, "I am so hungry I could eat the hide off of a mule."

Second Confederate soldier, "Well, as much as that thought is disturbing to me, I would join you if I could" says his comrade. We are in a perilous situation here. And we have been in it for time. If the good people of this community have not been as forthcoming as they have with their food we would be in a worse situation than the present one. But this army has always experienced food shortages. And I suspect that that situation will continue."

First Confederate soldier, "You're right. Why...at Shiloh in April of '62 my cousin, who fought there, told me that the men were so hungry that they could not fight the second day. Of course, that might just be an excuse for their poor performance on the second day, especially after they won the first day and failed to take advantage of their gains on the second. Of course, the Yankees had something to do with that. But it gets you to thinking."

Executive Mansion, Washington, D. C.

January 26, 1863

> To: Major General Joseph Hooker

> General:

"I have placed you at the head of the Army of the Potomac. Of course I have done this upon what appear to me to be sufficient reasons. And yet I think it best for you to know that there are some things in regard to which, I am not quite satisfied with you. I believe you to be a brave and a skillful soldier, which, of course, I like. I also believe you do not mix politics with your profession, in which you are right. You have confidence in yourself, which is a valuable, if not indispensable quality. You are ambitious, which, within reasonable bounds, does good rather than harm. But I think that during Gen. Burnside's command of the Army, you have taken counsel of your ambition, and thwarted him as much as you could, in which you did a great wrong to the country, and to a most meritorious and honorable brother officer. I have heard, of you recently saying that both the Army and the Government needed a Dictator. Of course, it was not for this, but in spite of it, that I have given you the command of the Army of the Potomac. Bu I remind you that only those generals who gain success, can set up dictatorships. What I now ask of you is military success, and I will risk the dictatorship. The government will support you to the utmost of its ability, which is neither more nor less than it has done and will do for all commanders. I much fear that the spirit that you have aided to infuse into the Army, of criticizing their Commander, and withholding confidence from him, will now turn upon you.

I shall assist you as far as I can, to put it down. Neither you, nor Napoleon, if he were alive again, could get any good out of an army, while such a spirit prevails in it. And now, beware of rashness, but with energy, and sleepless vigilance, go forward, and give us victories.

Yours very truly,
A. Lincoln

The President's faith in his new commander proved unwarranted. There would be no dictatorship established on his watch. Hooker's leadership during the next major battle, the Battle of Chancellorsville, would also prove to be severely wanting.

The Battle of Chancellorsville was the largest mismatch in strength and size among convened armies during the course of the entire war, pitting the largest federal army ever assembled, the Army of the Potomac, against its smaller rival the Army of Northern Virginia; the latter faced overwhelming odds. Yet, due to the leadership of the Army of Northern Virginia the smaller confederate army of 60,000 soldiers employing an improvised strategy in which Robert E. Lee divided his army, and in so bold a move, defeated the larger force of Major General Joseph Hooker's 130,000 men.

The Food Scams, Hardtack

The demands of large armies calls for many forms of support, among them is the call for food. And in this war, among the sparse food available to the soldiers of both armies was a commodity called hardtack, a unique mixture of flour, baking soda, salt and water that sustained the men in the field. It was barely a source of nourishment, and certainly not recommended from a culinary perspective. But nevertheless, hardtack was a durable, easy to prepare and transportable commodity that both armies could rely upon to offer sustenance even in the mist of the most dire circumstances. As for its other unique properties, hardtack was a food (resembling bread in a disguised form, one would be pressed to admit) that was likely to keep a modern dentists' occupied with bridge and upper plate work, and just as unlikely to satisfy a culinary or discerning palate. Its major benefit, among those listed was its portability and its ability to stay "fresh" for long periods of time.

Recipe for the Preparation of Hardtack

- 5 Cups Flour (unbleached)
- 1 Tablespoon Baking Powder
- 1 Tablespoon Salt
- 1-1 1/4 cups Water
- Preheated Oven to 450

Directions:

In a bowl, combine each of the listed ingredients to form a stiff, but not dry dough. The dough should be pliable, but not stick a lot to your hands.

Take this mound of dough, and flatten it out onto a greased cookie sheet (the ones with a small lip around the edge...like a real shallow pan...), and roll the dough into a flat sheet approx. 1/2 inch thick.

Using a bread knife, divide the dough into 3x3 squares. Using a 10-penny nail, put a 3x3 matrix of holes into the surface of the dough, all the way through, at even intervals.

Bake in the oven for approximately 20 minutes until lightly browned. Take out and let cool.

Do this the day before your go in the field, and you will have enough tack to fill your haversack. It will be somewhat soft on Saturday morning, but, by Sunday, you should soak it in your coffee before eating, else you will have a hard time chewing.

From: the *US Army Book of Recipes* (1862)

The Coffee Scheme

An army travels on its stomach. And food is a critical component affecting how well or poorly it functions. Among it's the most important offerings during the war in addition to hardtack was coffee that served both sides as a beverage, and as a restorative.

Coffee Warnings

While coffee was a prized commodity for many, this was not the case for all. As one opponent of its widespread adoption by the armies wrote:

"Don't use the stuff. There isn't one cook in five hundred who ever did anything else than abuse it. Some of the papers are recommending substitute- parched beans, rye, bread crusts, acorns, etc. Swamp mud will blacken water just as effectually, but neither of will make coffee... coffee fills your stomach with mud banks and shoals, against which the bark of human life is often wrecked. The greatest humbug in the world... is coffee! Think of paying forty cents a pound for charcoal to embitter and blacken the water you drink. The practice should be suppressed by the Board of Health, if there were no war to do it.
Quoted in the *Wilmington Daily Journal*, 3 October 1861.

Coffee Testament

Just as a few condemned the widespread use of coffee there were others that praised its value across even the most harrowing of circumstances. As this soldier attests:

"What a Godsend it seemed to us at times! How often after being completely jaded by a night march . . . have I had a wash, if there was water to be had, made and drunk my pint or so of coffee and felt as fresh and invigorated as if just arisen from a night's sound sleep!"
Union Army soldier John Billings, a veteran of the 10th Massachusetts Volunteer Artillery

Quartermaster and commissary duties dealing with getting food and equipment to soldiers in the field were some of the Army's most challenging tasks. Large volumes of edibles and expendables had to be purchased, packaged, stored, and transported in an organized manner under rarely predicable and often, incredibly difficult conditions. In organized settings, such as in places where regimental gatherings occurred this task was feasible. At the farthest end of the logistical chain, however, huddled around campfires in the field rain or shine, were individual soldiers needing to be clothed, armed, and especially, fed.

According to U.S. Army regulations, a soldier was entitled to receive: Twelve ounces of pork or bacon, or 1 pound 4 ounces of salt or fresh beef; 1 pound 6 ounces of soft bread or flour or 1 pound of hard bread, or 1 pound 4 ounces of corn meal. Each company of 100 soldiers to share among themselves 1 peck of beans or peas; 10 pounds of rice or hominy; 10 pounds of green coffee or 8 pounds of roasted and ground, or 1 pound 8 ounces of tea; 15 pounds of sugar; 1 pound 4 ounces of candles; 4 pounds of soap; 2 quarts of salt; 4 quarts of molasses.

Coffee was among commissary items most consistently issued and available to Federal soldiers each day. Coffee was a stimulant and supplied limited nutrition needed to sustain physical activity. It was also an easy commodity to store and transport. Correspondingly, it was also an ideal product for profiteers to manipulate and control.

Coffee was a "vital" commodity during the War for both sides. Often Armies traveled long distances with little more than hardtack and coffee to sustain them. Coffee, moreover, soon became a desirable trade commodity for both sides. Northern soldiers would trade coffee available to them, while Southern soldiers would swap tobacco in exchange for the beverage. Under these circumstances, where money was difficult to value, it soon became evident to profiteers that coffee was a most desirable and profitable commodity.

"For the stimulating property to which both tea and coffee owe their chief value, there is unfortunately no substitute; the best we can do is to dilute the little stocks which still remain, and cheat the palate, if we cannot deceive the nerves."
"Substitutes for Coffee," The Southern Banner, 1865

Cornering the Profiteers

Officials of the Army of the Potomac took keen interest in the trade of soldiers, especially where profit was a major factor among those seeking to gain advantage. As we listen in, we find Dr. Letterman and John in discussion. This time the topic is food.

"Food and clothing are essential parts of this war," says Dr. Letterman. "Better provisions will not only sustain our troops but also help them focus on the essentials of soldering. But, there are some troubling developments that we need to address. We are not operating at peak proficiency. Our supplies dwindle too soon and often appear to be of poor quality. We could use some help in improving the feeding of our troops. Clearly, there is a problem with our food distribution system that I would like you to look into. We have enough food, but it is not being properly prepared, and may be of poor quality judging by the increase in sick soldiers that we have seen. Moreover, the men complain of the inferiority of the meals that they have been eating. And, most recently, it has come to my attention that our supplies appear to run out sooner than before. Our shipments have not been revised so the source of this problem does not lie there. As with other facets of our supply train I am beginning to suspect theft. You have been a cook and have a good sense of how a kitchen such as ours needs to be run. Perhaps, you can look into this matter. You may be able to improve the situation. As such, I would appreciate any assistance that you could offer on this subject."

John is not surprised by Dr. Letterman's revelation.

"Well, Major, I can look into this problem", he responds. "However, you know my methods. I cannot make a direct inquiry. Thieves will close ranks if they are discovered and find a black man making inquiries of their activities. I would also face certain danger if my purposes were discovered. I will need some form of disguise to hide my intentions, with perhaps, some scenario that would lessen the impact of my presence. However, there might just be a simple way to meeting this problem. As we are about to enroll colored soldiers in the Army of the Potomac I wonder if we can use that as a pretext for my inquiries. Perhaps, if I am issued a sergeant's uniform and introduced as a supply officer of a newly formed black regiment I could gain entry into the food fraternity responsible for the purchase and distribution of food. If I can convince the perpetrators of this scam that I am a confederate seeking admission to their arrangement for profit, I may be able to accomplish this mission."

Early the next day John is introduced to Supply Sgt. Jim Tully who in turn, introduces him to Miles O'Connor a supplier of coffee and other products to the Army. O'Connor, a man rumored to be involved in questionable food purchases, may be the link that John is seeking in the profiteering food market scams. This projection is double-edged. O'Connor, ever open to making a quick profit, sees in John a new market and the prospect of making money. He decides to offer him a lucrative proposition.

"I have been informed Sgt. that we will be supplying your regiment with vittles. And we will try to get you the best fare that we can. That is, within the limits imposed by the scarcities produced by the war", he suggests with the wink of an eye. As John detects, he has stumbled into one of the food scams suggested by Dr. Letterman; there is obviously something irregular going on and O'Connor and Tully appears to be several of its perpetrators.

Playing to the proposition that appears to be emerging from this discussion, John responds by saying, "I assume that there will be a little left over for my boys and myself in any such transaction. I am personally addicted to coffee."

"Well Sgt. I think that we can accommodate you within our system of distribution. As a matter of fact, the complexity of the system of food distribution, that is simpler than assumed, would probably guarantee that event."

At that moment John withdraws from the pocket of his coat on official looking order prepared in advance with the assistance of Dr. Letterman, directing him to procure a small sample of food stores for his regiment. Among the items listed is coffee.

John then spying his interest, hands Connor the list of food supplies he has carefully prepared. Among the items listed is fifty pounds of coffee.

John proceeds by adding, "If you could provide me with the coffee now I will send some of my boys over to collect the rest of our provisions this afternoon."

"We can do that, says O'Connor. "However, there are some things that you should note that makes our distribution system work for all parties," says O'Connor. We have two "brands" of coffee here.* The one in the sacks marked by the green binding contains all coffee. The ones with the red binding contain coffee mixed with less costly ingredients we have been able to supply complementing the original product without affecting the taste much. There is no reason why you and your boys couldn't share the green sack and distribute the red sacks, or sell them to interested parties as we do. That is where our profit lies."

Herein lies the deception, the fraud that has been sought and suspected. John looks suspicious at O'Connor upon his play on words recognizing it for the scam that it is. But O'Connor sensing doubt is quick to respond to any of John's misgivings. He says, "There is no reason to doubt the coffee, nor will there be by others who consume it. The red sacks contain a blend that will still taste like coffee, but it maybe a bit weaker and somewhat bitter than that contained in the green sacks. We use this method of mixing products to extend our profit. And we have done so without much complaint. If you agree to this arrangement, as I am sure that you will, I am willing to offer you 10 percent of our profits in this food sharing enterprise. I might add that there are other arrangements that you could participate in as we get going." John now has the evidence he seeks, and seeing his opportunity to terminate the scheme, John agrees to the suggested arrangement.

Later that afternoon John sends 5 men and a wagon to collect the prepared food packet. While the transaction is being made and the wagon is being loaded three armed men on horseback enter the commissary. They are members of General Hancock's newly formed food inspector's squad tasked with overseeing food procurement for the army. As they enter the scene they call attention to O'Connor and his colleagues and despite a vigorous protest, they take them into custody for processing. For now, this end of the food scheme has been shut down.

*A number of traders, makers of ground coffee, in particular, would increase the amount of "fillers" they would add to their products. Unfortunately, there was no inspection of procured products or standards to be applied at the time. Subsequently, non-harmful fillers included corn, as well as barley, wheat, soybeans, rice and brown sugar were often added to the mixture constituting a somewhat different product than originally procured.

Aside from the obvious deception of false advertising and mismanagement, the coffee scam held the potential for causing harm among its consumers. Unfortunately, often there were also harmful fillers such as wood, twigs, husk and even dirt applied by unscrupulous traders. These fillers would be virtually impossible for consumers to detect once the coffee beans have been roasted and ground.

Southern War Scams

The economic disparities between the North and the South offered different opportunities for fraud. The North, being more prosperous and diverse in its ability to support varied enterprises offered potential war profiteers greater prospects in the procurement, manufacture and distribution of goods and services for mischief, some bordering on human suffering in order to make a profit. That state of affairs was true of the South, as well as the North. And among the vital products subject to manipulation and enormous profit was salt.

Before the Civil War, people in the South consumed an estimated 450 million pounds of salt annually. However, most of this vital product was not produced in the region; it came from Wales on ships, which carried salt as ballast when they sailed to Southern ports to pick up cotton. In the nineteenth century, salt was used in a variety of food and commercial products, including the processing of leather goods such as harnesses, belts and shoes. Salt's most important use, however, was as a preservative. In an age without refrigeration, virtually all pork and beef that was not cooked and served immediately after slaughter needed to be preserved for future consumption. In this manner salt was used to preserve fish, and meats. Salt was also used in cooking and was added as a condiment at the table. By mid-century the affection for salt grew enormously; Americans consumed more salt than any other nation in the world and among its consumers salt was used more often in southern cooking and in the South in general than in any other region of the United States.

Once the Union blockade was declared, and effectively enforced, ships no longer brought salt into Southern ports. New Orleans had large stockpiles of salt, but this accumulation had shrunk considerably by the fall of 1861. As a result of this action the price for salt increased substantially, reaching levels that many farmers who raised hogs were not able afford in order to preserve byproducts after the slaughtering of their animals because they had no salt to sustain their labor. In one instance, a Mississippi farmer wrote to the governor: "With a great many now, the deepest anxiety prevails to keep our families from suffering for want of salted provisions. Meat is now ready to be slaughtered."

The shortage of imported salt was only one reason for its rising price; another cause was speculation. As an editor of a Mississippi newspaper reported in November 1861 "all the salt in New Orleans and elsewhere is now in the hands of speculators. Something must be done in the matter, and be done quickly. We are willing that speculators should reap a rich profit, but we are not willing for them to suck the very life blood out of the people, if we can avoid it."

By 1862 the salt famine became severe. In March of that year an Alabama official reported that speculators were using "every artifice and fraud" to acquire salt. In May of 1862 the editor of Atlanta's Southern Confederacy told his subscribers "we will be in a dreadful condition unless we get salt." In December of that same year twenty women from Greenville, Alabama, became fed up with the salt famine. They marched on the local railroad station shouting "Salt or Blood," and forced an agent to give up the contents of a large sack of salt. Like the bread riots in Richmond the scarcity of salt provoked public outrage.

The Battle of Chancellorsville brought a mixed blessing of changes. After the discovery of the food scheme, measures were taken to reduce corruption and improve the soldiers diet. On the medical front the last component of the Letterman System now was in place: a new Army diet that included greater accountability, detailed cooking instructions and holding commissary officers accountable for insuring that the men received fresh food were put in place.

The results of these improvements in diet and nutrition were dramatic. Disease rates declined by half, and far fewer soldiers were being shipped north (and lost permanently to the war effort). The medical corps was now, as a function of these efforts, directly contributing to the sustained fighting strength of the Army. However, the consequences of the Battle itself proved disastrous.

An account of the dead and wounded involved in the Battle of Chancellorsville indicated that it cost more lives, and resulted in more casualties, than any other engagement ever fought on Virginia soil. Greatest among the losses suffered was the accidental, though mortal wounding of Gen. Thomas "Stonewall" Jackson by his own men on a dark night in which Jackson was surveying the battlefield at night.

Yet, because of the daunting odds that Gen. Robert E. Lee faced, and the dazzling tactical initiatives that he employed, Chancellorsville has been called his greatest victory. The toll in killed and wounded, however, would normally question any assessment of merit in terms of victory alone.

15

THE QUESTION OF SLAVERY

Prelude to Gettysburg

Sometime before the initiation of the war, in the presidential year of 1860, distressing news was transmitted to the people of the South.

On Friday morning, Nov. 9, 1860, Keziah Goodwyn Hopkins Brevard, a wealthy 57-year-old widowed plantation mistress, who owned over 200 slaves and who lived some 10 miles east of Columbia, SC, wrote in her diary,

"Oh My God!!! This morning heard that Lincoln was elected." In the hurried entry that followed, she recorded her thoughts and fears:

"I had prayed that God would thwart his election in some way & I prayed for my Country—Lord we know not what is to be the result of this—but I do pray if there is to be a crisis—that we all lay down our lives sooner than free our slaves in our midst—no soul on this earth is more willing for justice than I am, but the idea of being mixed up with free blacks is horrid!! I must trust in God that he will not forget us, as unworthy as we are—Lord save us—I would give my life to save my country. I have never been opposed to giving up slavery if we could send them out of our country—I have often wished I had been born in just such a country—with all our religious privileges & liberties with none of them in our midst—if the North had let us alone—the Master & the servant were happy with out advantages — but we had had vile wretches ever making the restless worse than they would have been & from my experience my own negroes are as happy as I am—happier. I never am cross to my servants without cause & they give me impudence if I find the least fault, this is of the women, the men are not half as impudent as the women are. I have left a serious & what has been an all absorbing theme to a common one but the die is cast — "Caesar has past the Rubicon." We now have to act. God be with us is my prayer & let us all be willing to die rather than free our slaves in their present uncivilized state."

On this date, her country was still the United States of America, but as an astute woman, she knew what Lincoln's election portended. It was in the air, and in the news, as well as on every body's lips. The next day, the South Carolina legislature called for a convention to consider secession, and on Dec. 20 the state seceded. "I wish Lincoln & Hamlin could have died before this & saved our country dissolution," Brevard confessed. As her male neighbors, she would support secession in order to defend her way of life.

The meaning of events can be and are often indiscernible following no recognizable script. However, there are also occasions when the simplest discussion reveals a spark of truth in response to a perplexing question. Such is the case with this interchange of two soldiers questioning the motivation of Southern troops engaged in battle.

Fred, a Northern soldier: "In my discussions with some of the captured Confederate soldiers they have told me that this war is being fought so that they can maintain their rights, that they are afraid that the right to own slaves would be stripped from them. But at far as I can tell their concern for their rights is not the real reason why they are here. It seems to me that slavery is the real issue."

Martin, his companion: "I think that you are right. That might have been part of the reason why many of these boys were here initially. At least that's what some of the veterans seem to believe, as I have been told. But it is not the cause of this conflict. Why most of them do not even own a single slave, nor are they likely to. Why I have heard tell that many of the Southern boys, deceived by this simple explanation and distortion of the truth, refer to this conflict as a "rich man's war and a poor man's fight.

That would imply some truth to the saying, and the fact that many of these boys have been tricked into serving a cause that is not their own."

The condemnation of the practice of slavery was echoed through many voices. One opinion was vociferously expressed by David Walker. He wrote in an antebellum tract, "It will be recollected, that in the first edition of my 'Appeal,' I promised to demonstrate to the satisfaction of the most incredulous mind, that we Coloured People of these United States, are the most wretched, degraded and abject set of beings that ever lived since the world began, down to the present day, and, that, the white Christians of America, who hold us in slavery, (or, more properly speaking, pretenders to Christianity,) treat us more cruel and barbarous than any Heathen nation did any people whom it had subjected, or reduced to the same condition, that the Americans (who are, not withstanding, looking for the Millennial day) have us. All I ask is, for a candid and careful perusal of this the third and last edition of my Appeal, where the world may see that we, the Blacks or Coloured People, are treated more cruel by the white Christians of America, than devils themselves ever treated a set of men, women and children on this earth."
David Walker (1830) *Appeal to the Colored Citizens of the World*, Boston.

The meaning of the war changed over time. Nevertheless, the question of slavery, its continuance and its spread to the new territories, was at its core from the beginning. A reading of the constitution of the newly formed Confederate States of America clearly testifies to this argument as it defines the cause of the war for us.

As stated in the Confederate Constitution, the place of slavery in the new country is well enunciated:

"No bill of attainder, ex post facto law, or law denying or impairing the right of property in negro slaves shall be passed."
Constitution of the Confederate States of America, Article I, Section 9, Clause 4.

Moreover, the constitution makes the case for slavery doubly clear as it goes on to state:

"The citizens of each State shall be entitled to all the privileges and immunities of citizens in the several States; and shall have the right of transit and sojourn in any State of this Confederacy, with their slaves and other property; and the right of property in said slaves shall not be thereby impaired."
Constitution of the Confederate States of America,
Article IV, Section 2, Clause 1

These admonitions did not go unchallenged. Anti-slavery writers attempted to offer counter arguments in the abolitionists' fight against slavery. Using books, newspapers, pamphlets, poetry, published sermons, and other forms of literature, abolitionists spread their message long before and during the war. They offered a moral appeal to others regarding the foul, degrading practice of slavery, a practice that had characterized America from its earliest beginnings until the end of the Civil War.

David Walker's Appeal, William Lloyd Garrison's The Liberator, and Frederick Douglass' The North Star were among the most important abolitionist writings. And then there were the slave narratives – the personal accounts of what it was like to live in bondage by those who suffered its debasing effects. Their stories would give the Northern people their closest look at slavery yet and provide an undeniable counter to the pro-slavery arguments and idyllic pictures of slavery described by slaveholders as appeasement to the practice.

One such story appeared in the autobiography of John P. Parker, former slave and later, conductor on the Underground Railroad.

During the period of his enslavement John Parker was assigned to a post under the direction of a harsh and abusive taskmaster. He writes, "My apprenticeship was of short duration. While I was willing, the man was a drunkard and began abusing me on the very first day. Being apt and observing, I soon could do rough jobs. Setting me to a task, which I knew was beyond my ability I did the best I could. The plasterer when he saw my work flew into a rage, [and] beat me with a lath with a nail in it, until I had to go to a hospital for slaves. It was kept by a white woman who was inexperienced, and a heartless creature, as she not only neglected her patients, but would beat them unmercifully at the least provocation.

I stood by helplessly and watched her beat the helpless. She was beating a woman with a rawhide whip when I protested. Instantly [she] struck me across the face. Without a thought of what I was doing, I seized the whip and gave the white woman a sound beating, then ran out of the house, knowing full well what would happen to me if I was caught. All day long I hid in the piles of freight on the New Orleans boat dock, determined I would take the boat that night. I had been to New Orleans and knew my way about; that was as far as I could think. One thing I did know: I had to get out of Mobile.

...I stole on board the New Orleans steamer, and launched myself on an adventure that carried [me] into strange places and stranger incidents." John P. Parker, *His Promised Land*

Northern abolitionists compiled and wrote of the denigrations suffered by members of the slave population. And, above all they preached and wrote of the hopelessness of the lives of the slaves. Many of their writings appeared during the War. However, the retched condition of the slave population was addressed earlier, as well. T. Weld (1839) along with the Grimke sisters, in one of the earliest attacks on the practice of slavery wrote, "Look at the slave, his condition but little, if at all, better than that of the brute; chained down by the law, and the will of his master; and every avenue closed against relief; and the names of those who plead for him, cast out as evil;-must not humanity let its voice be heard, and tell Israel their transgressions and Judah their sins?

"May God look upon their afflictions, and deliver them from their cruel task-masters! I verily believe he will, if there be any efficacy in prayer. I have been to their prayer meetings and with them offered prayer in their behalf. I have heard some of them in their huts before day-light praying in their simple broken language, telling their heavenly Father of their trials in the following and similar language."

A Slave's Account

"Fader in heaven, look upon de poor slave, dat have to work all de day long, dat cant have de time to pray only in de night, and den massa mus not know it.*

"Fader, have mercy on massa and missus. Fader, when shall poor slave get through de world! when will death come, and de poor slave go to heaven;" and in their meetings they frequently add, "Fader, bless de white man dat come to hear de slave pray, bless his family," and so on. They uniformly begin their meetings by singing the following..."

Weld continues, "Every man knows that slavery is a curse. Whoever denies this his lips libel his heart. Try him; clank the chains in his ears, and tell him they are for him; give him an hour to prepare his wife and children for a life of slavery; bid him make haste and get ready their necks for the yoke, and their wrists for the coffle chains, then look at his pale lips and trembling knees, and you have nature's testimony against slavery.

"Two millions seven hundred thousand persons in these States are in this condition. They were made slaves and are held such by force, and by being put in fear, and this for no crime! Reader, what have you to say of such treatment? Is it right, just, benevolent? Suppose I should seize you, rob you of your liberty, drive you into the field, and make you work without pay as long as you live, would that be justice and kindness, or monstrous injustice and cruelty? Now, every body knows that the slaveholders do these things to the slaves every day, and yet it is stoutly affirmed that they treat them well and kindly, and that their tender regard for their slaves restrains the masters from inflicting cruelties upon them. We shall go into no metaphysics to show the absurdity of this pretense. The man who robs you every day, is, forsooth, quite too tender-hearted ever to cuff or kick you! True, he can snatch your money, but he does it gently lest he should hurt you. He can empty your pockets without qualms, but if your stomach is empty, it cuts him to the quick. He can make you work a life time without pay, but loves you too well to let you go hungry. He fleeces you of your rights with a relish, but is shocked if you work bareheaded in summer, or in winter without warm stockings. He can make you go without your liberty, but never without a shirt. He can crush, in you, all hope of bettering your condition, by vowing that you shall die his slave, but though he can coolly torture your feelings, he is too compassionate to lacerate your back--he can break your heart, but he is very tender of your skin. He can strip you of all protection and thus expose you to all outrages, but if you are exposed to the weather, half clad and half sheltered, how yearn his tender bowels! What! Slave holders talk of treating men well, and yet not only rob them of all they get, and as fast as they get it, but rob them of themselves, also; their very hands and feet, all their muscles, and limbs, and senses, their bodies and minds, their time and liberty and earnings, their free speech and rights of conscience, their right to acquire knowledge, and property, and reputation;--and yet they, who plunder them of all these, would fain make us believe that their soft hearts ooze out so lovingly toward their slaves that they always keep them well housed and well clad, never push them too hard in the field, never make their dear backs smart, nor let their dear stomachs get empty.

"But there is no end to these absurdities. Are slaveholders dunces, or do they take all the rest of the world to be, that they think to bandage our eyes with such thin gauzes? Protesting their kind regard for those whom they hourly plunder of all they have and all they get! What! when they have seized their victims, and annihilated all their rights, still claim to be the special guardians of their happiness! Plunderers of their liberty, yet the careful suppliers of their wants? Robbers of their earnings, yet watchful sentinels round their interests, and kind providers for their comfort? Filching all their time, yet granting generous donations for rest and sleep? Stealing the use of their muscles, yet thoughtful of their ease? Putting them under drivers, yet careful that they are not hard-pushed? Too humane forsooth to stint the stomachs of their slaves, yet force their minds to starve, and brandish over them pains and penalties, if they dare to reach forth for the smallest crumb of knowledge, even a letter of the alphabet!" Weld, T. Grimke, A. & Grinke, S. (1839), *Slavery as It Is*

The slave narratives were immensely popular with the public. Frederick Douglass' Narrative of the *Life of Frederick Douglass* sold 30,000 copies between 1845 and 1860, while William Wells Brown's Narrative went through four editions in its first year, and Solomon Northup's *Twelve Years a Slave* sold 27,000 copies during its first two years in print. Many narratives, such as these, were translated into other languages, wetting the appetites and curiosity of foreign reader including the British, French, German, Dutch and Russian consumer.

In addition to publishing their narratives, former slaves became anti-slavery lecturers and went on tour in order to share their experiences. They told their stories to audiences throughout the North and in Europe. Frederick Douglass was the most famous of these author-lectures. However, he was not alone in his recitations. He was joined by others among them Sojourner Truth and William Wells Brown. In addition, many others, such as Ellen and William Craft -- a couple who had escaped together using ingenious disguises -- lectured but did not create a written narrative. For white audiences who had perhaps never seen an African American man or woman, the effects of these articulate people telling their stories was electrifying and won many supporters and adherents to the abolitionist cause.

Some former slaves, such as Douglass and Brown, wrote their narratives themselves. However, many ex-slaves denied the opportunity to learn to write and read were illiterate. These ill-educated people, however, were not doomed to be forgotten forever; they dictated their stories to sympathetic others who wrote of their trials and hardships under slavery.

One self-composed experience of slavery lived by a person who bore witness to the evils of the institution, as described here in multiple forms, was offered by the ex-slave Frederick Douglass.

"The reader will have noticed that among the names of slaves, Esther is mentioned. This was a young woman who possessed that which was ever a curse to the slave girl - namely, personal beauty. She was tall, light-colored, well formed, and made a fine appearance. Esther was courted by "Ned Roberts," the son of a favorite slave of Col. Lloyd, who was as fine-looking a young man as Esther was a woman. Some slave-holders would have been glad to have promoted the marriage of two such persons, but for some reason, Captain Anthony disapproved of their courtship. He strictly ordered her to quit the company of young Roberts, telling her that he would punish her severely if he ever found her again in his company. But it was impossible to keep this couple apart. Meet they would, and meet they did. Had Mr. Anthony been himself a man of honor, his motives in this matter might have appeared more favorably. As it was, they appeared as abhorrent as they were contemptible. It was one of the damning characteristics of slavery, that it robbed its victims of every earthly incentive to a holy life. The fear of God and the hope of heaven were sufficient to sustain many slave women amidst the snares and dangers of their strange lot; but they were ever at the mercy of the power, passion, and the caprice of their owners. Slavery provided no means for the honorable perpetuation of the race. Yet despite of this destitution there were many men and women among the slaves who were true and faithful to each other through life.

But for Esther, that was not to be the case. "Abhorred and circumvented as he was, Captain Anthony, having the power, was determined on revenge. I happened to see its shocking execution, and shall never forget the scene. It was early in the morning, when all was still, and before any of the family in the house or kitchen had risen. I was, in fact, awakened by the heart-rending shrieks and piteous cries of poor Esther. My sleeping-place was on the dirt floor of a little rough closet that opened into the kitchen, and through the cracks in its unplanned boards I could distinctly see and hear what was going on, without being seen. Esther's wrists were firmly tied, and the twisted rope was fastened to a strong iron staple in a heavy wooden beam above, near the fire-place. Here she stood on a bench, her arms tightly drawn above her head. Her back and shoulders were perfectly bare. Behind her stood old master, with cowhide in hand, pursuing his barbarous work with all manner of harsh, coarse, and tantalizing epithets. He was cruelly deliberate, and protracted the torture as one who was delighted with the agony of his victim. Again and again he drew the hateful scourge through his hand, adjusting it with a view of dealing the most pain-giving blow his strength and skill could inflict. Poor Esther had never before been severely whipped. Her shoulders were plump and tender. Each blow, vigorously laid on, brought screams from her as well as blood. "Have mercy! Oh, mercy!" she cried. "I won't do so no more." But her piercing cries seemed only to increase his fury. The whole scene, with all its attendants, was revolting and shocking to the last degree, and when the motives for the brutal castigation are known, language has no power to convey a just sense of its dreadful criminality. After laying on I dare not say how many stripes, old master untied his suffering victim. When let down she could scarcely stand. From my heart I pitied her, and child as I was, and new to such scenes, the shock was tremendous. I was terrified, hushed, stunned, and bewildered. The scene

here described was often repeated, for Edward and Esther continued to meet, not withstanding all efforts to prevent their meeting."
From: *The Autobiography of Frederick Douglass*

The slave narratives provided the most powerful voices contradicting the slaveholders' propaganda and favorable claims of charity and generosity that they exhibit toward their slave captives. By their very existence, the narratives demonstrated that African Americans were people with a mastery of language and the ability to write or relate to others in oral testament their own history. The narratives, moreover, told of the horrors of family separation, the sexual abuse of black women, and the inhuman workload that they were called upon to bear. They also told of free blacks being kidnapped and sold into slavery. They described the frequency and brutality of floggings and the severe living conditions of slave life. They also told exciting tales of escape, heroism, betrayal, and tragedy. The narratives captivated readers, portraying the fugitives as sympathetic, fascinating characters.

The narratives also gave Northerners a glimpse into the life of slave communities: the love between family members, the respect for elders, and the bonds between friends. They described an enduring, truly African American culture, which was expressed through music, folktales, and religion. Then, as now, the narratives of ex-slaves provided the world with its closest look at the lives of enslaved African American men, women and children. They were the abolitionist movement's voice of reality.

While the slave narratives were immensely popular, the anti-slavery document that would reach the broadest audience was written by a white woman. She was Harriet Beecher Stowe. Stowe was less threatening to white audiences than were black ex-slaves. Her anti-slavery message, moreover, came in the form of a novel, a popular vehicle that was even more accessible to a wide audience. It was called *Uncle Tom's Cabin*.

Harriet Beecher Stowe, though not an active abolitionist herself, had strong anti-slavery feelings. She had grown up in an abolitionist household that had harbored fugitive slaves. She had also spent time observing slavery first-hand on visits to Kentucky, across the river from her Cincinnati home. With the passage of the Fugitive Slave Act of 1850, Stowe decided to make a strong statement against the institution of slavery. She had been working as a freelance journalist to supplement her husband's small income and help support their six children. In June 1851 Stowe began publishing *Uncle Tom's Cabin* in serialized form in the *National Era*.

The response to her work was extraordinarily enthusiastic, and people clamored for Stowe to publish the work in book form. It was a risky business to write or publish an anti-slavery novel in those days, but after a great deal of effort she found a reluctant publisher. In a cautious gesture, only 5,000 copies of the first edition were printed. Yet, the response was overwhelming and this limited edition was sold out in two days. The demand, moreover, increased substantially over time. By the end of the first year, 300,000 copies had been sold in America alone while in England 200,000 copies were sold, including one read by the queen of England. In response to its popularity the book was translated into numerous languages and was also adapted for the theater in many different versions, becoming a theatrical success that played to enthusiastic audiences throughout the world.

Uncle Tom's Cabin had a tremendous impact on its reading public. The character Uncle Tom is an African American who retains his integrity and refuses to betray his fellow slaves at the cost of his life. His firm Christian principles in the face of his brutal treatment made him a hero to whites. In contrast, his tormentor Simon Legree, the Southern slave-dealer turned plantation owner, enraged them with his cruelty. Stowe convinced readers that the institution of slavery itself was evil, because it supported people like Legree and enslaved people like Uncle Tom. Because of her work, thousands rallied to the anti-slavery cause.

Southerners, however, predictably were outraged by Stowe's novel, and declared the work to be criminal, slanderous, and utterly false. Among the reactions in the South, a bookseller in Mobile, Alabama, was forced out of town for selling copies, while Stowe received threatening letters and on one occasion, a package containing the dismembered ear of a black person. Southerners also reacted by writing their own novels. These depicted the happy lives of slaves, and often contrasted them with the miserable existences of Northern white workers laboring under depressing working conditions.

Most black Americans responded enthusiastically to *Uncle Tom's Cabin* including, among them Frederick Douglass who was a friend of the author. Stowe had consulted him on some sections of the book, and he commended the book in his writings. At the time, most black abolitionists saw the book as a tremendous help to their cause. Some, however, critically opposed it, seeing Uncle Tom's character as being weak and ineffectual, too submissive and compliant and criticized Stowe for having her strongest black characters immigrate to Liberia.

16

THE BATTLE OF GETTYSBURG

The Battle of Gettysburg, Pennsylvania (July 1–July 3, 1863), was the largest battle ever fought on American soil, involving near 85,000 men in the Union's Army of the Potomac under Major General George Gordon Meade and approximately 75,000 men in the Confederacy's Army of Northern Virginia, commanded by General Robert Edward Lee. The toil of the battle was enormous; casualties at Gettysburg totaled 23,049 for the Union (3,155 dead, 14,529 wounded, 5,365 missing) while Confederate casualties consisted of more than 28,063 (3,903 dead, 18,735 injured, and 5,425 missing), more than a third of Lee's army.

These largely irreplaceable losses to the South's largest army, combined with the Confederate surrender of Vicksburg, Mississippi, on July 4, marked what is widely regarded as a turning point—perhaps the turning point—in the Civil War, although the conflict would continue for nearly two more years of destruction with witness to several more major battles, including Chickamauga, Spotsylvania Courthouse, Mononacy, Nashville, Petersburg and others.

The Gettysburg Campaign: The Fighting Begins

In the wake of the Confederate victory at Chancellorsville, Virginia (May 1–4, 1863), Robert E. Lee decided to attempt a second invasion of the North. This would take pressure off Virginia's farms during the growing season, especially in the "breadbasket of the Confederacy," the Shenandoah Valley. In addition, any victories won on Northern soil would put political pressure on Abraham Lincoln's administration to negotiate a settlement to the war, or might lead to the South's long hoped-for military alliance with England and France.

The campaign began under a dark shadow: Lee's creative and aggressive corps commander, Lieutenant General Thomas "Stonewall" Jackson, previously had been mortally wounded by his own men at Chancellorsville in an accidental occurrence several months earlier. Consequently, the Army of Northern Virginia was reorganized from two corps to three, with two new commanders, Lt. Gen. Richard "Dick" Ewell replacing Jackson in the Second Corps and Lt. Gen. Ambrose Powell (A. P.) Hill selected to command the newly formed Third Corps. Lieutenant General James Longstreet—Lee's "Old War Horse"— would retain command of the First Corps. The Army of Northern Virginia was about to invade enemy territory with two of its three corps commanders newly appointed to their positions. As could be predicted, the secretive, self-reliant Jackson had done little to prepare them for this level of command.

This would be Lee's second incursion into the North. The previous one ended in the bloodiest single day in America's history with the Battle of Antietam (called the Battle of Sharpsburg in the South) in Maryland on September 17, 1862. Total casualties from that one-day battle exceeded 23,000.

In order to mask the army's movement up the Shenandoah Valley into western Maryland and central Pennsylvania, Lee depended upon his renowned cavalry leader J.E.B. "Jeb" Stuart. Upon crossing into Maryland, Stuart loosely interpreted Lee's poorly formed, ambiguous orders and began raiding Union supply trains. Cut off by the advancing Army of the Potomac, from June 25 until the night of July 2, Stuart lost all communication with the rest of the Confederate army, leaving Lee to operate blindly deep within enemy territory.

Meanwhile, on the Union side, the Army of the Potomac was still under the command of General Joe Hooker, who had lost the Chancellorsville battle, diminishing his reputation as "Fighting Joe" through apprehensiveness in the face of the enemy. As reports arrived that the

Confederates had crossed the Potomac and were on Northern soil, Hooker dispersed his army widely, trying to simultaneously protect the approaches to Washington, Philadelphia and Baltimore. He'd lost Lincoln's confidence, however, and the President made the difficult, as well as daring decision to replace his chief army commander in the face of an enemy invasion. On June 28, a military engineer, Major General George Gordon Meade, a nonpolitical general who had only been promoted to corps command less than six months earlier was placed in charge of the Union's largest army. This surprise appointment was not Lincoln's first choice; yet, the general accepted it. Meade was a reluctant, but experienced leader. However, he had under his command, experienced and talented lieutenants. And they advised him well. Under their counsel Meade ordered his scattered corps to concentrate on a mountain ridge south of the town of Gettysburg. In this manner his army could each quickly reinforce another in the event of an attack. He hoped to draw Lee into attacking him on high ground along Pipe Steam Creek.

As Meade's corps moved closer to each other, Lee's army was scattered, moving along multiple roads. Lee had issued orders to his subordinates to not bring about a general engagement until the army could concentrate its forces. Fate, however, intervened with other plans.

"About three weeks before the battle, rumors were again rife of the coming of the rebel horde into our own fair and prosperous State. This caused the greatest alarm; and our hearts often throbbed with fear and trembling. To many of us, such a visit meant destruction of home, property and perhaps life.

"We were informed they had crossed the State line, then were at Chambersburg, then at Carlisle, then at or near Harrisburg, and would soon have possession of our capital. We had often heard of them taking horses and cattle, as well as carrying off property and destroying buildings.

"A week had hardly elapsed when another alarm beset us.

"The Rebels are coming! The Rebels are coming!" was passed from lip to lip, and all was again consternation."

Tillie Pierce

Excepts drawn from the impressions penned by Tillie Pierce, a young resident of the town, on her observations of the events that took place in Gettysburg, Pennsylvania in July of the summer of 1863.

Gettysburg: Day 1

On the morning of July 1, Major General Daniel Henry Heth of A.P. Hill's Third Corps, sent his 7,500-man division down the Chambersburg Pike toward Gettysburg. Encountering resistance, they initially assumed it was more of the hastily assembled Pennsylvania Emergency Militia that they'd been skirmishing with during the campaign.

In reality, Colonel John Buford had deployed part of two brigades of Union cavalry as skirmishers in the brush along Willoughby's Run three miles west of town. Just two weeks previously, his troops been issued rapid fire breech-loading carbines, and they used the guns' fast-loading capability to create the impression of a much larger force, slowing the advance of Hill's brigades for a time before falling back.

The Confederates followed them across the stream, only to meet a line of Union infantry on McPherson's Ridge. The Army of the Potomac was arriving piecemeal, and among the first to arrive was a brigade of Western regiments that had earned the nickname the "Iron Brigade of the West." The Confederates recognized these "fellows in the black hats" as they were known, and realized, from the well-earned reputation of their adversaries that they were in for a rougher day than expected.

Union Major General John Reynolds, the well-regarded competent commander of the left wing of the Army of the Potomac (I, III and XI Corps), arrived and took charge of the defense. His men fought tenaciously, but Reynolds was shot sometime during the initiation of the battle by a Confederate sharpshooter, and died from his wounds.

Learning of Reynold's death, Meade acted quickly. From his headquarters at Taneytown, Meade dispatched Major General Winfield Scott Hancock to take command at Gettysburg, although Major General O. O. Howard was already on the field. He was to assess whether or not the battle should be fought there. Hancock, seeing the strong defensive position offered by the hills south of Gettysburg, chose to stand, and Meade ordered the other Corps to the little crossroads town.

By afternoon, Confederate reinforcements had also arrived, and the general engagement Lee hadn't wanted at this stage of the campaign was a fait accompli.

The Union's XI Corps was driven back through the town of Gettysburg, losing 4,000 men, and by evening was entrenched in a defensive posture on Culp's and Cemetery hills south of town.

Lee expressed a desire for General Ewell to assault the hills without waiting for further reinforcement, but he failed to make it an expressed order. Ewell, recognized that his men were exhausted from the day's fighting, and being unfamiliar with Lee's directions, did not press his tired soldiers forward, giving Meade time to reinforce the troops on the hills.

Gettysburg: Day 2

James Longstreet's corps had arrived, and on the second day his 20,000 men were sent to outflank the Union left, which was anchored to the south by two hills known as Little Round Top and Big Round Top. Lee, seizing an opportunity, had learned from an early morning reconnaissance report that the Federals had failed to place troops upon those hills. Ewell, meanwhile was ordered to make a demonstration against Culp's and Cemetery hills on the Union right flank and to use his own discretion about launching a full-scale attack.

Longstreet's men, moving toward their objective, had to reverse, countermarch and take a different route after Brigadier General Lafayette McLaws discovered the planned route would put them in full view of the Federals, negating any advantage of surprise. This cost valuable time but, as events turned out, a Union general was about to present them an unexpected opportunity.

All but one of Meade's seven corps were now on the field, deployed in an inverted fish-hook shape with its center along Cemetery Ridge; the defensive positions on Culp's and Cemetery hills formed the hook at one end in order to allow the Commanding General to shift troops from either end of the line to the center as need dictated.

The left was assigned to Major General Daniel Edgar Sickles, a non-professional, nor a trained soldier who owed his military rank to his political importance in the essential state of New York.

Dissatisfied with his position at the lower end of Cemetery Ridge, (Sickles had suffered a previous defeat under similar circumstances.) Sickles took it upon himself (without consultation, as would be appropriate) to advance his III Corps nearly a half-mile west toward the Emmitsburg Pike and open high ground in a wheat field near a peach orchard. The move dangerously stretched his 10,000-man Corps. Sizing the opening provided by Sickles's, Longstreet's men attacked Sickles's new position, and the fighting at rocky Devil's Den, the wheat field and the peach orchard was among the fiercest and bloodiest of the three days.

Meade faced with Sickles' blunder, sent the V Corps and part of the XI to reinforce him. But they would not be able to rectify Sickles's mistake. Earlier in the day New York's Irish Brigade received last rites from a Catholic priest before charging into the fray; it was a well-intentioned gesture since 198 of them would not return from the desperate fighting in the hot, sultry afternoon.

17

CHAMBERLAIN AT GETTYSBURG

Above the blood-soaked fields, a similar drama was playing out on Little Round Top, the Federals anchor on the left. Around 4:30 p.m., the men of Alabama, Texas and Arkansas regiments, from John Bell Hood's Division in Longstreet's Corps, began ascending the steep hill from the west. Had they arrived two hours earlier, they would have captured the nearly deserted heights unopposed. However, by the time they arrived Meade's chief of engineers, Brig. Gen. Gouveneur K. Warren, had discovered the unoccupied position on the extreme left of Cemetery Ridge; if fallen to the enemy it would present a potentially irreversible disastrous situation. Warren sent messages to Sickles requesting reinforcement. However, his own situation precluded him sending reinforcements as that time.

One message, nevertheless, found its way to Colonel Strong Vincent, commanding the 3rd Brigade, 1st Division, of the Federals' V Corps. He double-timed his men and deployed them among the rocks and trees of Little Round Top's western and southern slopes. The fate of the Union Army, at that moment, rested on the shoulders of 1,350 men of the 83rd Pennsylvania, 44th New York, 16th Michigan and 20th Maine regiments. Vincent's orders were to "hold this ground at all costs!"

The Union defensive line on aptly named Cemetery Ridge resembled an inverted fishhook, extending from Culp's Hill on the north, down

Cemetery Ridge and southward toward Big and Little Round Tops. Although the 650-foot-high Little Round Top was overshadowed by its larger neighbor its position was critical; much of the hill was cleared of trees and it could better accommodate troops. Strategically, Little Round Top held the key to the developing battle. If the Southern troops could take and hold the hill, they could theoretically roll up the entire Union line.

JOSHUA LAWRENCE CHAMBERLAIN 'S MEDAL OF HONOR

The above Congressional Medal of Honor was awarded to Chamberlain in 1893 for his actions at Gettysburg, Pennsylvania, on July 2, 1863. Joshua Chamberlain's medal was donated anonymously to the Penobscot Historical Society in Brunswick Maine. It had followed an unusual legacy. Earlier it had been donated by Chamberlain's estate to a church in Duxbury, Massachusetts where the donor unknowingly had purchased it during a church book sale.

The 34-year-old Chamberlain was one of the most interesting figures in the Civil War. A highly cultured, somewhat sedentary professor-scholar of modern languages at Maine's exclusive Bowdoin College, he had sat out the first year of the war on Bowdoin's campus. But in July 1862, sensing perhaps that the war was going to last a good deal longer than he had first believed Chamberlain offered his services to the Union cause. "I have always been interested in military matters", he informed Maine Governor Israel Washburn, "and what I do not know in that line, I know how to learn." He was given command of the newly formed 20th Maine, a unit comprised of extra men left over from other newly formed regiments. It was not, Chamberlain noted, one of the state's favorite fighting units — No county claimed it; no city gave it a flag; there was no parade held in its honor and there was no send-off at the station.

The 20th Maine had been organized under President Abraham Lincoln's second call for troops on July 2, 1862. The regiment initially fielded a total complement of 1,621 men, but by the time of the Battle of Gettysburg, especially after the Battle of Fredericksburg, the stress of campaigning had reduced the regiment's ranks to some 266 soldiers; the 20th Maine was considered a weak link in Vincent's brigade. Fortune, in a disguised form, however, was to smile on Chamberlain's regiment in the form of unexpected reinforcements.

On May 23, 1863, 120 three-year enlistees from the 2nd Maine Infantry were marched under guard into the regimental area of the 20th Maine. The 2nd Maine men were in a state of mutiny and refused to fight, angry because the bulk of the regiment—men with only two-year enlistments — had been discharged and sent home, and the regiment had been disbanded, while these soldiers remained on the field. The mutineers claimed they had only enlisted to fight under the 2nd Maine flag, and if their flag went home, so should they. By law, however, the men still owed the Army another year of service.

Chamberlain had orders to shoot the mutineers if they refused to fight. Fortunately for the men of the 2nd Maine, a confluence of circumstances prevented that occurrence. Chamberlain was born and grew up in Brewer, Maine the twin city to Bangor across the Penobscot River where the 2nd Maine regiment was recruited. The mutineers were not just anonymous soldiers but were also Chamberlain's childhood neighbors; he could not have them shot. Instead, Chamberlain wisely distributed the 2nd Maine veterans evenly to fill out the 20th Maine's ranks and integrated the experienced soldiers within the 20th Maine. He sympathized with the mutineers and promised them that he would write to Maine Governor Abner Coburn, after the battle in order to straighten out the mix-up in three-year versus two-year contracts that they had signed. On Little Round Top the 120 experienced combat veterans from the 2nd Maine brought the 20th's ranks up to 386 infantrymen and added manpower to help hold Chamberlain's wobbling line together.

As he arrived on Little Round Top, Colonel Strong Vincent chose a line of defense that started on the west slope of the hill. When the first regiments reached the rocky outcrops in that area, Vincent put them into line. The 16th Michigan took up a position on the right flank, and the 44th New York and 83rd Pennsylvania held the center. Later in life, Chamberlain wrote that his regiment was the first in line, but it actually took up its position last, curving its line back around to the east and forming the Union Army's extreme left flank.

The last thing Vincent would tell Chamberlain was: "This is the left of the Union line. You are to hold this ground at all costs!" Chamberlain ordered the regiment to go on line by file. He deployed Company B, recruited from Piscataquis County and commanded by level headed Captain Walter G. Morrill of Williamsburg, forward to the regiment's left front flank as skirmishers. Company B, with its 44 men, was subsequently cut off by a flanking attack by the enemy, leaving the 20th with only 314 armed men on the main regimental line.

Also helping to defend Little Round Top were Major Homer R. Stoughton's 2nd U.S. Sharpshooters, armed with .52-caliber breech loading rifles. These sharpshooters' skirmishing abilities were unequaled in the Union Army, and a 14-man squad was attached to Company B. The men took up a position in a ravine east of Little Round Top.

Shortly after the Federals had taken up their positions, the 824 men of the 4th and 5th Texas regiments of Maj. Gen. John B. Hood's division hammered up the slope of Little Round Top, pushing toward the center and right of Vincent's line. During that assault, Captain James H. Nichols, the commander of the 20th Maine's Company K, ran to alert Chamberlain that the Confederates seemed to be extending their line toward the regiment's left. Chamberlain called his company commanders together and told them his battle plans. With the new information from Nichols,

Chamberlain ordered a right-angle formation, extending his line farther to the east.

Meanwhile, Colonel Vincent tried to rally his 3rd Brigade as the 16th Michigan staggered under the heavy assault by the 4th and 5th Texas. Just when the Federals were on the verge of collapse, Colonel Patrick O'Rourke led the 140th New York Zouaves into the gap to save Vincent's brigade. Unfortunately, as the onslaught increased in ferocity, both Vincent and O'Rourke paid with their lives for their heroism.

Elements of Hood's division, the 15th and 47th Alabama, then began to smash into the Maine troops. Hood ordered these regiments, led by Colonel William C. Oates, to find the Union left, turn it and capture Little Round Top.

Twenty-five-year-old Color Sgt. Andrew J. Tozier of the 2nd Maine quickly emerged as an unlikely hero; he was later awarded the Medal of Honor for his bravery. It had been Chamberlain's idea to elevate Tozier to the post of color sergeant for the 20th Maine, a move designed to instill a new esprit de corps in the mutineers. Color sergeant was a dangerous but coveted position in Civil War regiments, generally manned by the bravest soldier in the unit. As the 20th Maine's center began to break and give ground in the face of the Alabama regiments' onslaught, Tozier stood firm, remaining upright as Southern bullets buzzed and snapped in the air around him. Tozier's personal gallantry in defending the 20th Maine's colors became the regimental rallying point for Companies D, E and F to retake the center. Were it not for Tozier's heroic stand, the 20th Maine would likely have been beaten at that decisive point in the battle.

When their ammunition had almost run out, Chamberlain decided upon an improbable course of action; it was an unanticipated and unexpected move. His men, out of ammunition, were ordered to fix bayonets and charge down into the two Alabama regiments. Chamberlain later said he communicated his decision to counterattack to Captain Ellis Spear, the acting battalion commander of the unit's left flank. Spear, however, claimed he received no such orders.

Corporal Elisha Coan, a member of the 20th Maine's color guard, claimed that 1st Lt. Holman S. Melcher, the acting commander of Company F, actually conceived the idea to advance the colors and that

Colonel Chamberlain initially hesitated, fearing that it would be extremely hazardous. Coan said other officers joined Melcher in urging a forward movement.

Chamberlain, whose right foot had been pieced by a shell fragment or a stone chip, then limped along the regimental line giving instructions to align the left side of the regiment with the right. After Chamberlain returned to the regimental center, Melcher asked permission to retrieve his wounded from the front. Chamberlain replied, "Yes, I am about to order a right wheel forward of the whole regiment." (Chamberlain himself claimed later to have said, yes, in a moment! I am about to order a charge.)

Chamberlain then ordered a right-wheel maneuver and took up a place behind Tozier. There is some disagreement about exactly what Chamberlain said as a bayonet charge was ordered. One story is that he screamed: "Bayonet! Forward to the right!" Chamberlain claimed later that one word — "Bayonet!" — was enough and that it would have been in vain to order his troops forward because no one could hear it over the noise generated by the sounds of the ensuing battle. Nor was there time. Right wheel or Bayonet! Forward to the right was perhaps someone's post-war idea of what Chamberlain would have said if time permitted. The state-appointed Maine commission that later gathered facts regarding Maine's contribution to the Battle of Gettysburg maintained that Melcher sprang forward as Chamberlain yelled, Bayonet! and that Chamberlain himself was abreast of the colors.

With all the confusion and noise on Little Round Top that day, if anything other than bayonet had been said it probably would not have mattered, anyway. Many of the infantry were now out of ammunition, and faced being cut down on the next enemy assault. Upon hearing the metal-to-metal sound of bayonets being put on en masse, each man would know the intent of the upcoming order without actually hearing it. In all likelihood Lieutenant Melcher conceived the idea to advance the colors to retrieve the wounded, but Chamberlain expanded upon the idea, deciding to have the whole regiment conduct a bayonet attack.

After Chamberlain ordered Bayonet! the Union line hesitated until Melcher sprang out in front of the line with his sword flashing. Captain

Spear said he never received a formal order to charge — he charged only after he saw the colors start forward.

Later, Chamberlain wrote of the battle:

"The roar of all this tumult reached us on the left and heightened the intensity of our resolve. Meanwhile the flanking column worked around to our left and joined those before us in a fierce assault, which lasted with increasing fury for an intense hour. The two lines met and broke and intermingled in the shock. The crush of musketry gave way to cuts and thrusts, grapplings and wrestlings. The edge of conflict swayed to and fro, with wild pools and eddies. At times I saw around me more of the enemy than of my own men; gaps opening, swallowing, closing again with sharp, convulsive energy; squads of stalwart men who had cut their way through us, disappearing as if translated all around me, strange, mingled roar- shouts of defiance, rally and desperation; and underneath, murmered (sic) entreaty and stifled moans; gasping prayers, snatches of Sabbath song, whispers of loved names; everywhere men torn and broken, staggering, creeping, quivering on the earth, and dead faces with strangely (sic) fixed eyes staring stark into the sky.

"In the very deepest of the struggle while our shattered line had pressed the enemy well below their first point of contact... I saw through a sudden rift in the thick smoke our colors standing alone. I first thought some optical illusion imposed upon me. But as forms emerged through the drifting smoke, the truth came to view. The cross fire had cut keenly; the center had almost been shot away; only two of the color guard had been left, and they fighting to fill the whole space; and in the center, wreathed in battle smoke, stood the Color Sergeant Andrew Tozier. His color-staff planted in the ground at his side, the upper part clasped in his elbow, so holding the flag upright, with musket and cartridges seized from the fallen comrade at his side he was defending his sacred trust in the manner of the songs of chivalry. It was a stirring picture..."

During the charge, a second enemy line consisting of the 15th and the 47th Alabama tried to make a stand in the proximity of a stone wall. For a moment it looked as though the Confederates might succeed in halting the Federals and breaking their momentum. But, using the classic element of surprise, Captain Morrill's Company B rose up from behind an adjacent stone wall and fired a volley into the Confederates' rear, breaking the will of the enemy troops. Confederate reports showed that the Union company had been magnified into two regiments. According to Confederate Colonel Oates, it was the surprise fire of Company B that caused the disastrous panic in his soldiers.

Chamberlain may have been blessed with both good timing and good luck. He not only had made the right command decisions but also had survived his attack. An Alabama soldier facing him in battle twice failed to pull the trigger of his rifle because he had second thoughts about killing the brave colonel. Then a pistol aimed and fired by a Southern officer misfired only a few feet from Chamberlain's face.

Without the stand of Sergeant Tozier inspiring others to close up and bolster the sagging middle of the regiment, the Confederate attacks could have eliminated the 20th Maine as a fighting force. Tozier's bravery sparked the 20th Maine and changed the course of the engagement. Without Tozier, there would not have been an opportunity for Chamberlain to attack.

LITTLE ROUND TOP

By the time the sun went down on the second day at Gettysburg, the Union left still held, but the III Corps would no longer be a significant factor in the battle, and the V Corps had been badly mauled. Meanwhile, a desperate contest was taking place on the slope of Cemetery Hill.

On Culp's Hill

On July 2, 1863 fighting occurred throughout the course of the day. Apart from the left, other activity was taking place on the Union right flank. Here they prepared for a Confederate assault by Ewell's troops who were now advancing from the town of Gettysburg in an assault upon Culp's and Cemetery hills. The going was difficult. For an hour the Confederates struggled across rough ground while Union batteries threw shot and shell among them. But when they got far enough up the slopes, to overcome the Federals the Rebels routed the infantry of the XI Corps. As Union regiments pulled from one area of Cemetery Hill to plug a gap created by the retreat they created their own gap, and the Confederate infantry in response poured through the opening.

Down on Cemetery Ridge, Winfield Scott Hancock sent the 14th Indiana and 7th West Virginia regiments to reinforce Cemetery Hill. Arriving after dark, they formed up and charged into the Rebels who were fighting with artillerymen around the Union guns. At this juncture The Confederates fell back. In one of the ironic events of the war, the 7th West Virginia, which had been the 7th Virginia until June 20 when West Virginia was admitted as a free state, fought hand to hand with the 7th Virginia of the Confederacy, in the process capturing a nephew of their own regimental commander.

The long day of bloodshed finally ended. At midnight Meade called together his commanders for a council of war. He'd already sent a message to the War Department stating that he intended to stay and fight. Meade's army had been attacked on the left and on the right; that fact, combined with other intelligence he'd received, led him to believe his center would be the target the next day.

18

Day 3: Pickett's Charge

Throughout the war, Robert E. Lee had always sought a way to "get at those people over there." His aggressiveness had served the Confederate cause well on many battlefields, but on July 3, 1863, it led to disaster.

An Unheeded Warning:

Major General James Longstreet and Commanding General Robert E. Lee had fought many successful battles together. However, on this day the former presented a dissenting argument, contrary to the nature of their relationship, that determined the course of this pivotal engagement.

Many years after the war, Longstreet wrote: "Feeling that he was still in his disposition to attack, I tried to anticipate him, by saying: General, I have had my scouts out all night, and find that you still have an excellent opportunity to move around to the right of Meade's army, and maneuver him into attacking us."

He replied with his fist, "The enemy is there, and I am going to strike him."

I felt that it was my duty to express my conviction; I said General, I have been a soldier all my life. I have been with soldiers engaged in fights by couples, by squads, companies, regiments, divisions, and armies, and I should know, as well as anyone, what soldiers can do. It is my opinion that no fifteen thousand men ever arrayed for battle can take that position," pointing to Cemetery Hill.

General Lee, in reply to this ordered me to prepare Pickett's Division for the attack. I should not have been so urgent had I not foreseen the hopelessness of the proposed assault. I felt that I must say a word against the sacrifice of my men; and then I felt that my record was such that General Lee would or could not misconstrue my motives. I said no more, however, but turned away."

Despite the passionate arguments of his second in command James Longstreet that a charge toward the Union center was designed to fail, Lee instructed his "Old War Horse" to strike the Union center on Cemetery Ridge, using the divisions of Brig. Gen. James Johnston Pettigrew, Maj. Gen. Isaac Ridgeway Trimble, and the recently arrived division of Maj. Gen. George Pickett. In all, approximately 15,000 men were to advance three-quarters of a mile across open ground, climb fences along the roads, and charge up the gradual but steep slope of Cemetery Ridge to assail a force of about 6,500.

At 1:00 in the afternoon, a prolonged artillery barrage by the Confederates, utilizing an unprecedented number of guns spread two miles wide, preceded the assault, intended to silence the Union's cannons and weaken the infantry. Amidst the significant haze and smoke caused by this massive barrage, however, most of its shells over shot their intended target and went high, plunging to earth behind the Federals' line. The Army of Northern Virginia spent its most lethal supply of ammunition in a wasted effort. The federals remained in place.

Initially, Federal guns replied to this attack, until the order came down to conserve ammunition for the attack that was obviously coming. When the Union cannons fell silent, Lee's artillery chief, Col. Edward Porter Alexander, sent word for Longstreet to bring up his men.

Pettigrew's division of four brigades formed the left of the attack line, with two of Trimble's brigades behind them and to their right for support. Pickett's men stepped out on the right.

The magnificent line, initially formed on the onset of the march, however, deteriorated as the men moved forward and the massive artillery took its toil. The advance was disordered by terrain and by flanking fire on

Pettigrew's left as it neared the Union line. Pickett's advance drifted left, exposing his right to enemy fire. Through shot, shell, canister and rifle fire, the long Confederate line surged forward. Near the Union center, it broke through temporarily until federal reinforcements drove it back. As the survivors straggled back to Confederate lines at Seminary Ridge, many of them passed Robert E. Lee, who told them, "It is my fault."

Aftermath

On July 4, Lee started a 27-mile-long train of hospital wagons down the road to Virginia. His army halted at the flooded Potomac River and entrenched for another battle, but Meade's army, was battered and exhausted and had consumed much of its ammunition. The Army of the Potomac did not pursue Lee, for which Meade would be soundly criticized later. Meade was made subordinate to Grant where he remained in command of the army for the rest of the war. Grant, was promoted to lieutenant general and placed over all Northern armies; Grant forthwith attached himself to the Army of the Potomac where he directed the Army of the Potomac for the remainder of the war. Lee offered his resignation to Confederate president Jefferson Davis, but it was refused and he, too, remained in command of his Army for the rest of the war.

One soldier's memorial to the battle.

Here at the foot of the mountain the engagement became general and fierce and lasted until 8 o'clock at night. And in the third and last charge the fatal blow was struck.

My Brother: You have offered your life as a sacrifice
upon your country's altar.
Today concludes the term of life of my bother.
He now sleeps upon the battlefield of Gettysburg with
There brothers, fathers, small & great,
Partake the same repose
There in peace the ashes mix
Of those who once were foes.

Robert Ware, soldier, CSA

19

THE WOUNDED AND DYING

At Gettysburg, the Union 11th Corps ambulance train consisted of 100 ambulances, nine medical wagons, 270 men, and 260 horses.

Lieutenant John S. Sullivan of the 14th Indiana observed Letterman and other medical personnel during the thick of the action at Gettysburg. His observations were recorded in the following note:

"And many a time did I see the stretcher-carriers fired upon and wounded while bearing away the wounded. But they did not desist from their humane work; and many a time did I watch anxiously, fearing every moment to see him fall. He coolly rode all over the field, sometimes in the thickest of the firing, and away to the front even of our pickets on his errand of mercy, not satisfied to leave a single suffering man uncared for on the bloody field. All honor to such noble fellows."

Several residents of the town recalled their observations of the Battle of Gettysburg, as well. Isaac Carter, who had spent the battle hiding in a woodlot for fear of being captured, recounted those events for us:

"I visited the battlefield three days after the fight, and it made me sick the bodies were so numerous and so swelled up, and some so shot to pieces – a foot here, an arm there, and a head in another place. They lay so thick in the Valley of Death that you couldn't walk on the ground. Their flesh was black as your hat – yes, black as the blackest colored person."

A SILENT CANNON

During the Battle of Gettysburg in July of 1863, Letterman established a General Hospital and temporary field hospitals wherever there was a source of water and shelter. The buildings used ranged from churches, farm buildings and private homes. Sometimes the only shelter available was trees or a piece of canvas strung between poles. As the battle approached its end, there were approximately 23,000 wounded from both the Union and Confederate armies to deal with. Fortunately, via Letterman's intervention his system saved the lives of many of these men.

The following is an excerpt from US Army Medical staff:

THE MEDICAL AND SURGICAL HISTORY
OF THE WAR OF THE REBELLION
CONTAINING REPORTS OF MEDICAL DIRECTORS,
AND OTHER DOCUMENTS

Edited, under the direction of Surgeon General Joseph K Barnes, United States Army, by Assistant Surgeon J. J. Woodward, United States Army, and Assistant Surgeon George A. Otis, United States Army

Washington, D.C.: 1870

"Over six hundred and fifty medical officers are reported as present for duty at that battle. These officers were engaged assiduously, day and night, with little rest, until the 6th, and in the Second Corps, until the 7th of July, in attendance upon the wounded. The labor performed by these officers was immense. Some of them fainted from exhaustion induced by over exertion, and others became ill from the same cause. The skill and devotion shown by the medical officers of this army were worthy of all commendation; they could not be surpassed. Their conduct as officers and as professional men was admirable. Thirteen of them were wounded; one of whom, Surgeon W. S. Moore, 61st Ohio, Eleventh Corps, died on the 6th of July, from the effects of his wounds received on the 3d. The idea, very prevalent, that medical officers are not exposed to fire, is thus shown to be wholly erroneous.

The greater portion of the surgical labor was performed before the army left. The time for primary operations had passed, and what remained to be done was to attend to making the men comfortable, dress their wounds and perform such secondary operations as from time to time might be necessary.

One hundred and six medical officers were left behind when the army left; no more could be left, as it was expected that another battle would, within three or four days, take place; and, in all probability, as many wounded be thrown upon our hands as at the battle of the 2d and 3d, which had just occurred. I asked the Surgeon General, July 7th, to send twenty medical officers to report to Surgeon H. Janes, hoping they might prove of some benefit, tinder the direction of the medical officers of this army who had been left behind. I cannot learn that they were ever sent. Dr. Janes was left in general charge of the hospitals, and, to provide against contingencies, was directed, if he could not communicate with me, to do so directly with the Surgeon General, so that be had full power to call directly upon the Surgeon General to supply any want that might arise.

The ambulance corps throughout the army acted in the most commendable manner during those days of severe labor. Notwithstanding the great number of wounded, amounting to fourteen thousand one hundred and ninety-three, I know, from the most reliable authority and from my own observation, that not one wounded man of all that number was left on the field within our lines early on the morning of the 4th of July. A few were found after daylight beyond our farthest pickets, and these were brought in, although the ambulance men were fired upon, when engaged in this duty, by the enemy, who were within easy range. In addition to this duty, the line of battle was of such a character, resembling somewhat a horse-shoe, that it became necessary to remove most of the hospitals further to the rear, as the enemy's fire drew nearer. This corps did not escape unhurt: one officer and four privates were killed, and seventeen wounded, while in the discharge of their duties. A number of horses were killed and wounded, and some ambulances injured. These facts will show the commendable and efficient manner in which the duties devolving upon this corps were performed, and great credit is deservedly due to the officers and men for their praiseworthy conduct. I know of no battle-field from which wounded men have been so speedily and so carefully removed, and I have every reason to feel satisfied that their duties could not have been performed better or more fearlessly.

Before the army left Gettysburg, and knowing that the wounded had been brought in from the field, six ambulances and four wagons were ordered to be left from each corps to convey the wounded from their hospitals to the railroad depot for transportation to other hospitals. From the Cavalry Corps, but four ambulances were ordered, as this corps had a number captured by the enemy at or near Hanover a few days previously. I was informed by General Ingalls that the railroad to Gettysburg would be in operation on the 6th, and upon this based my action. Had such been the case, this number would have been sufficient. As it proved that this was not in good running order for some time after that date, it would have been better to have left more ambulances. I acted, however, on the best information that could be obtained.

The number of our wounded, from the most reliable information at my command, amounted to fourteen thousand one hundred and ninety-three. The number of Confederate wounded who fell into our hands was six thousand eight hundred and two; making the total number of wounded thrown by that battle upon this department twenty thousand nine hundred and ninety-five. The wounded of the 1st of July fell into the hands of the enemy, and came under our control on the 4th of that month. Instruments and medical supplies belonging to the First and Eleventh Corps were in some instances taken from the medical officers of those corps by the enemy."

The magnitude and destruction caused by the three-day battle of Gettysburg was beyond anyone's imagination. Reports from the battlefield on the number of killed, wounded and missing were staggering. No battle on American soil saw so many persons fallen or in need of care.

20

OBSERVERS AND HELPERS

Let us return to the beginning of the battle (July 1, 1863) and the medical mission it commanded. As we enter the scene on this first day of the battle Major Jonathan Letterman is supervising the removal of the dead and wounded in the vicinity. Accompanying him is John Scobell. They are in the midst of discussing the consequences of the opening salvo of the battle.

"Mr. Scobell the devastation visited upon this community is enormous. And it will continue", I'm afraid. It is way beyond the capacity of the normal resources of the town to address our needs here. We will require way stations and places where the wounded can be treated. Can you help in this effort?

"Well Sir! I will try with the assistance of your orderlies to locate places for refuge among the citizens for fallen soldiers. And I also suspect that other places, temporary way stations will be needed, as well. When the fighting dies down, perhaps we can build some interim quarters that will serve us, as well. Fortunately, with this hot weather we need not concern ourselves with sheltering the men from snow or frost as at Fredericksburg and we can erect outdoor shelters of a simpler design. That should require fewer resources and expedite the matter more quickly and smoothly. But we should be prepared for some rough times even if circumstances improve."

The efforts by John and his assistants were now underway. And there were many men assigned to the tasks of that day and during those that followed. Most of the transitional "hospitals" created involved the requisitioning and use of erected structures such as churches, schools and homes where they could be found. Among these efforts nearly 1,900 soldiers were carted off the Gettysburg battlefield and onto the George Spangler farm during the first days of July 1863.

In this case the work produced mixed results. What was not evident at first was that the Spangler farm was not a safe place to care for injured soldiers. Neither by design nor intent artillery fire would be over-arching that site and some of the soldiers stationed there would be fired upon based on their proximity to the farm. Artillery fire exacerbated an already agonizing scene and added to the suffering already exacted and continuing.

One General who toured the facility noted a common occurrence at such places, pools of blood, and amputated limbs lying in heaps. For six weeks the Spangler family was relegated to just one room, as wounded soldiers and staff occupied almost every other inch of the property. But the job that was initiated there was completed.

July 3: The surgeon's work

Another appropriated facility was the Weikert farm. From there teenager Tillie Pierce reported:

> "Toward the close of the afternoon it was noticed that the roar of the battle was subsiding, and after all had become quiet we started back to the Weikert home. As we drove along in the cool of the evening, we noticed that everywhere confusion prevailed. Fences were thrown down near and far; knapsacks, blankets and many other articles, lay scattered here and there. The whole country seemed filled with desolation.
>
> "Upon reaching the place I fairly shrank back aghast at the awful sight presented. The approaches were crowded with wounded, dying and dead. The air was filled with moanings, and groanings. As we passed on toward the house, we were

compelled to pick our steps in order that we might not tread on the prostrate bodies.

"When we entered the house we found it also completely filled with the wounded. We hardly knew what to do or where to go. They, however, removed most of the wounded, and thus after a while made room for the family.

"As soon as possible, we endeavored to make ourselves useful by rendering assistance in this heartrending state of affairs. I remember Mrs. Weikert went through the house, and after searching awhile, brought all the muslin and linen she could spare. This we tore into bandages and gave them to the surgeons, to bind up the poor soldier's wounds.

"By this time, amputating benches had been placed about the house. I must have become inured to seeing the terrors of battle, else I could hardly have gazed upon the scenes now presented. I was looking out of the windows facing the front yard. Near the basement door, and directly underneath the window I was at, stood one of these benches. I saw them lifting the poor men upon it, then the surgeons sawing and cutting off arms and legs, then again probing and picking bullets from the flesh.

"Some of the soldiers fairly begged to be taken next, so great was their suffering, and so anxious were they to obtain relief.

"I saw the surgeons hastily put a cattle horn over the mouths of the wounded ones, after they were placed upon the bench. At first I did not understand the meaning of this but upon inquiry, soon learned that this was their mode of administrating chloroform a chemical used in order to produce unconsciousness. But the effect in some instances was not produced; for I saw the wounded throwing themselves wildly about, and shrieking with pain while the operation was going on.

"To the south of the house, and just outside of the yard, I noticed a pile of limbs higher than the fence. It was a ghastly sight! Gazing upon these, too often the trophies of the amputating bench, I could have no other feeling, than that the whole scene was one of cruel butchery."

21

A FINAL RESCUE

The major cause of the war originally forged by Southern proponents was made by appeal to concern for States Rights and individual freedom. Paradoxically, this argument was made by both the stated purpose of the War, as well as the continuing presence of slavery in the existing slave states and its threatened expansion west into the territories. The existence of slavery itself as an institution defined much of the United States of America from its origins. And it was a long barbaric history.

From 1619 onward, before the actual formation of the country slavery was in place. The first known slave trade occurred in Jamestown Virginia. Of course, the trade did not end there. The speed and spread in the number of human beings selling and trading other human beings, based on the color of their skin, increased dramatically.

Initially used as farm laborers the slave population served other industries, as well. During the late antebellum era the use of hired slave laborers was one of the key elements in maintaining Virginia's economy. The contracting of slave labor hiring was already an established facet of Virginia's antebellum industries, where male slaves comprised a large portion of the workforce in iron factories and on railroad lines. This involvement of slaves in the economy, increased over time, often in the face of idle white men. And it continued.

During the war, private employers like the Tredegar ironworks, specializing in the production of war material in Richmond, as well as railroad lines, salt works, and iron forges, all of which sustained the Confederate war effort, hired increasing numbers of slave laborers as their white employees left for the army. Owners also leased their slaves to individual officers within the Confederate army or larger departments like the Confederate Medical Department, which hired hundreds of male and female slaves to work as nurses, cooks, and laundresses in army hospitals. Moreover, the war increased the importance of slaves with industrial skills in the upper South's hiring market; the demand for hired field hands also increased as white men joined the Confederate army in large numbers.

These labor initiatives were necessary to sustain the South's participation in the war. As confederate vice president Alexander H. Stephens of Georgia reported, slavery was the ideological "corner-stone" of the Confederate government. But it was slave labor that supported this philosophy. Slave labor provided the physical cornerstone for the Confederate war effort as well as its civilian foundation. Civilian and military employers in Virginia hired slaves in increasing numbers over the course of the war. Most of Virginia's slaves worked as agricultural laborers, and their wartime production helped feed both civilians and soldiers. Because the Confederacy's military and industrial employers typically only hired male slaves, much of the wartime agricultural work in Virginia was left to female slaves.

Paths to Freedom.

From its 17th century origins and throughout the succeeding centuries, black people have been kidnapped, transported and brought and sold to the highest bidders, thereafter to be held in perpetual bondage in the Americas. Their path to freedom, sporadic and often temporary, characterized much of their history in slavery. But this pattern began to change with the advent of the Civil War. Throughout the war the call for freedom now awakened the spirit of large numbers of runaway slaves who sought refuge in the Union army camps as Northern troops moved south. These runaway slaves predictably saw the Federal Army as a vehicle releasing them from slavery and allowing them to acquire their long sought freedom from the oppressive South and the institution of slavery. Among these men was a slave named John Brown who fled to the Union army, as others had done, where the Northern encampments and its army were seen as a gateway to this long desired dream.

In a letter to his wife, Brown wrote:

"My Dear Wife it is with grate joy I take to let you know Whare I am i am in Safety in the 14th Regiment of Brooklyn this Day I can Address you thank god as a free man I had a little truble in giting away But as the lord led the Children of Isrel to the land of Canon So he led me to a land Whare freedom Will rain in spite of earth and hell Dear you must make your Self content i am free from al the Slavers."

In his letter Brown also explained to his wife how he attained his freedom and the rapture he felt once he finally did. His actions, and those of his fellows made a difference. Slaves began undermining the South once they began running to the North. The Union army facilitated this by encouraging slave rebellions, while at the same time indicating to the Confederacy that the Union army would aid slave insurrections. These slave insurrections changed the balance of power on plantations, giving more control to the slaves that in turn, gave slaves the power to run away to fight for the Union army.

Joseph Miller: A Fugitive Runaway

Joseph Miller was a runaway. Like a number of others, among the many that shared his fate, he escaped the dehumanizing state of slavery that the color of his skin bound him. He was in his 42nd year when opportunity and necessity demanded that he flee from his condition of servitude.

It had been a lifelong dream that he shared with many of his brethren. His current situation, however, was somewhat complicated by the turn of events in which he found himself; while he was presently free from the binds of slavery, he was also a wanted man.

In the state of Virginia, where he was held captive for all of his life he had "murdered" a man, his master, and thereafter he was viewed as an outlaw and a fugitive from justice. Shortly after his encounter with the man who had wronged him, and his family for generations, he had fled from Virginia knowing that the possibility of his finding justice in a system rigged in favor of the slave owner was impossible. So one night, after an argument with his master Leon Miller, in which he had been threatened with the whip once again, a punishment that he had suffered many times in his adult life, he stabbed his owner with a purloined knife he had stolen from the kitchen where he was a servant. Now Joseph was on the run being pursued by his owner's sons, Prince and Simon whom sought to capture him and return with him to Virginia to stand trial for his crime and likely, a swift and sure death sentence. Had he not fled, he knew that his future would have been placed in jeopardy, a frequent occurrence in such cases. He knew that he could not remain in Virginia after the unfortunate incident and he fled for his life.

The story of Joseph's plight was far from unusual. He and his family had come to America as slaves as many other black people. Over time his descendants were bought and sold as the fortunes of their white masters necessitated. They were like the cattle in the field. As Joseph came of age he found a good slave woman who consented to be his wife and together they had five children. Among them was Jewel, a child of exceptional poise and beauty. However, she did not go unnoticed. At the age of thirteen, her master sent for her and misused her to his own satisfaction. Subsequently, she became pregnant with his child. However, she was too young to successfully bear the child, and who without the aid of an attending physician, died in childbirth with its mother. One week after the death of his daughter and granddaughter Joseph decided to confront Leon Miller. Joseph had a request to make of his master and he would do so.

"My daughter Jewel died in childbirth, a condition caused by you. You made her with child and refused to send for a doctor and get her proper care when she needed it most. She was just a piece of furniture to you, to be used and tossed aside. And you sir, as you and your kin can attest, have abused us, as well. As you personally have done with my daughter, just a child struck down in the prime of her life by you. There can be no justice in this matter. However, you can make amends by consenting to my request to grant my family a release from their present state of servitude."

Exasperated and astonished by this unusual request, Leon Miller turns to Joseph with a whip in his hand and spits out his vehemence saying, "I will not give up my property. Nor will I discuss this matter any further with the likes of you. Now I ask you to leave my presence before I take up the whip. That is my right and I will do so willingly." And as Leon Miller unravels his whip in a threatening manner, Joseph brandishes a knife stolen from the kitchen. Upon seeing the knife, Leon Miller backs away slowly and says, "Now put down that knife boy. You are in trouble and that knife will only add to your difficulties." But Joseph surprising his master by his defiance desists and Miller inches toward him. As Miller slowly approaches, Joseph detects his hot breath and lunges forward, proceeding toward him by inserting his blade into Leon Miller' stomach. Within moments, shocked by Joseph's act Leon reaches for his stomach and cries out in pain and shock. Meanwhile, Joseph realizes that he has committed an indefensible act for a black man. His only recourse at this time is to flee this place, knowing that his action will soon be discovered and he will be taken into custody by white authorities.

July 8, 1863

As we enter the scene we find John Scobell at the Letterman hospital compound. He is tenuously holding another man gently escorting him to a bed.

"Doctor, Doctor," cries John, "Come quickly. I have a man here that needs your help. Apparently, he has been shot and will require some attention."

"Who is it?" asks Letterman..

"I don't know," responds John. He appears to have wandered into camp after traveling some distance. By the looks of him he hasn't eaten for days. He also appears to have sustained, from those wounds that are apparent, a variety of bruises through rought living."

Upon hearing his request Dr. Letterman proceeds to attend to Joseph. After a physical investigation he suggests to John, "Well, aside from some minor aid and restoration to his body, I think that with proper nourishment and rest your man here will survive his ordeal.

"As others, he may also choose to join our cause and rid this nation of the abomination of slavery. Although I suspect that he has an interesting story to tell us, as well. And in due time Joseph does tell his story. Moreover, he relates his future to his saviors, as well.

"I cannot return. My future has been determined for me. I have been judged and sentenced, as any black man would be under the circumstances that I have described to you. I can protest, if given the chance, that I am innocent of their charges. But that is unlikely. I have been condemned already. I know the white people's system of justice in that part of the country. To return would be to confirm what I already know. If I am forced to go back to Virginia I swear to you that I will die trying to make my escape from this unjust and cruel fate that awaits me."

22

A Farewell to Arms

It is a bittersweet ending to our story as we enter a poignant scene in which our principal participants John Scobell and Dr. Jonathan Letterman stand astride a hill south of Gettysburg and observe the wheat field. They are standing together upon Little Round Top and are in the process of taking leave of each other. They now occupy an empty site, one upon which Joshua Chamberlain and his men successfully held the left flank of the Federal line one week earlier. Their mood, as one would suspect, is somewhat somber as they survey their surroundings. Letterman had just informed Scobell of his decision to leave the Army and to travel west.

"Come with me," he beckons his friend. "You have been part of this for some time and must be exhausted by what you have seen of this war. You can escape the conflict and the many deaths you have been witness to if you decide to go west. From what I hear, it is virgin country there, a place to start anew. Yes, they too have been touched by this war, as we all have been. But there is far less devastation there and fewer problems to cope with in that part of the country since it has been less exposed to this war and its destruction than there has been here."

"No, I cannot. The fate of my people lies here where this unfortunate conflict resides. I must help bring this horror to a conclusion, if I can. And that is here, in spite of the personal consequences that I may suffer.

There will come a time, probably at the war ends, when I will likely travel west. But that time is not now.

"But your life is worth less here than elsewhere. I appreciate your sentiments. But the fate of the black man is not tied to this region of the country. The black man has been treated unfairly in all parts of this country, north south, east and west. As you know, that has always been the case, and I suspect, that will continue to be so for some time to come. To run from this fact is self-delusional," adds the doctor.

"I need to stay and face this fact," says John. All black people must. And where necessary, we need to continue to fight for our rightful place in society. In this spirit I hope, white people of fair spirit and good heart will do, as well. Together we can overcome this unfortunate legacy. But that will only occur when we join hands together. Doctor, you are one of few white men that I have ever met that has treated me fairly and as a man. You have taught me that I have dignity and worth. I now need to call upon others of your race, and mine as well, many of who have have lost faith, for a more eventful life to do the same."

"But you have sacrificed so much, John. You have seen and done so much throughout this war. Others can finish the job now."

"Well, on that basis if I leave, I will be saying to all those who sacrificed their lives in this recent effort and previous battles, lives lost in trying to correct the injustices in our system of government, that they will have died in vain. Look around you doctor. Most of these men were young boys, often the first in their families to leave the place of their birth, many who could not tell you where or why they were here. Most, moreover, did not anticipate the fate to which they were doomed. I have learned that many of those seriously injured often seek death in preference to a life scarred by permanent, unanticipated wounds encountered through battle.

"I recently visited young Tim Collins, who lost his right arm on the third day of fighting. You may have saved his life. Yet, he will be going home to a life of hardship, assuming he gets that far. He will be a hero to some. Yet, he will no longer be able to plow his fields, nor hold his children or his wife, as he was once able. The shame is that he will be seen as only half a man by others, and most importantly, by himself. I cannot leave that legacy to lie here on this war-torn battlefield."

Civil War Amputations. Medicine on the battlefield during the Civil War was crude at times and good at best. With only a surgeon or two with an assistant surgeon to a regiment there was a good chance the wounded would severely overwhelm the medical personnel. There simply were not enough doctors to keep up with the rate of killing associated with the two armies. Moreover, the task of treating the wounded was equally compounded by the conditions under which physicians toiled.

The efficacy of treatment was predicated upon the type of injuries suffered on the battlefield, as well as the training and competence of those attending the wounded. This was especially complicated by the variety and disposition of gunshot and explosive wounds suffered by the men. Men, unfortunately, sustained many types of injuries, none more severe than, comminuted fractures, compound fractures, and wounds of the joints.

The difficulties in treating these wounds were often profound and intractable. The goal of the surgeon in many cases led to the necessity of amputation in order to provide the patient the best chance of survival.

Comminuted fractures consisted of those fractures where the bone is broken, splintered, or crushed into a number of different pieces. It does not take long to understand how a projectile, such as the minie ball, could produce such an injury, The minie ball, and other related projectiles that were fired from a rifle, were relatively slow moving and large. When the human body is struck at the bone the velocity of the projectile does not allow a clean exit. This produces a shattering effect at the point of impact. In general, when a comminuted fracture occurred with rupturing of the principal artery or nerve of a large limb it demanded amputation.

Compound Fractures are those fractures that produce a bone protruding from the skin. As in injuries producing a comminuted fracture the compound fracture may require amputation if sustained in a large limb, such as the leg, thigh, the arm, or forearm. Amputation during the Civil War was almost always performed if the large limb had severed principal arterial damage or produced nervous trunk damage.

The caring for those wounded on the battlefield varied across sites. Among the caregivers, Elizabeth Bliss Thacher Souder and many others rushed to offer their services to the wounded and the dying. In the following note she shares her experience of Gettysburg with her brother.

Gettysburg, Pennsylvania July 20, 1863

To: J. A. Thacher

My Dear Brother:

You will be surprised to receive a letter from me, bearing "Gettysburg" as a postmark. I left home, with a little company of friends, a week ago today, and arrived here after a wearisome journey, via Baltimore, on Tuesday night.

We commenced our labors at once in the field hospital of the Second Corps, to distribute milk punch, prepared from condensed milk, an invaluable thing in the hospitals, and to prepare nourishing food for our wounded soldiers, corn-starch and farina, eggs in various shapes, and nicely made tea and coffee. Each day as we have opportunity, we visit the soldiers in their tents, and try to speak a word of cheer. I was surprised and much interested to find the Colonel of the First Minnesota, with Lieutenant Mason and several of their regiment, in the same tent. I promised to take them under my special charge. Colonel Colvill was much pleased o find that I was your sister, and wanted to know if I intended writing to you soon. I will get the names of these Minnesota boys and send them to you, if possible. I was glad to be able to give to each of them an orange, a luxury much caved and difficult to obtain at this season, and took to them some chicken soup, which they thought very comfortable. I have seen a number of Minnesota boys, but have not their names, and am sorry to add, have seen also the graves of many Minnesotans.

A great many Maine boys are here, especially of the 19th Maine, which was terribly cut up.

We have several times visited the Adjutant of the 17th Maine, a pleasant young man from Portland, who bears the suffering from an amputated limb with great cheerfulness.

He is quartered in a private house in the town. On the opposite side of the street is a young captain of the same regiment, who had lost an arm, Captain Young. He was wounded, I believe, in the first day 's battle, and like many others laid several days in the woods without attention.

The faces of these New England boys light up with pleasure as they learn that several of our company are from the East, though now residents of the City of Brotherly Love, and all the soldiers love to hear of Philadelphia.

The beautiful fields around Gettysburg bear painful evidence of the recent struggle. We are very busy every day and have not attempted to visit any points of interest.

Wednesday morning – A man name Crowley, of the 1st Minnesota, who enlisted at St. Paul, was breathing his last, when we left camp last evening. Lieutenant M. will go to Baltimore today. He is an interesting young man, wounded in the hand. It is impossible to remember the names and identity of these soldiers, except in particular instances. We shall probably remain here another week, if we continue reasonably well. Colonel Colvill has several times asked if I had send my letter to you yet, and desired his regards. I will send it, therefore, without any further delay. The ambulance is waiting for us. Goodbye."

Your Affectionate Sister

WINSLOW HOMER – *Harper's Weekly*
ILLUSTRATION OF CIVIL WAR VETERAN AUGUST 26 1865

Like other wounds, especially those involving the extremities, those centering on the joints almost always required amputation during the Civil War. These wounds included injuries sustained to the knee joint, elbow joint, shoulder joint, wrist, ankle, and hip joint. Amputations made at these joints are also referred to as disarticulations. When an injury occurred at the arm below the forearm that required amputation, it was common to amputate through the elbow joint even though it was thought better by some to leave some of the radius and ulna. This was also the line of thinking when an amputation was required of the upper arm, leaving a few inches below the humerus. To prevent exposing the patient to inflammation and spread of infection upwards disarticulation was often performed.

As the discussion continues Dr. Letterman says to John, "I want to go as far as I can from this place, where this war has not yet touched so many men nor as deeply as it has done to us. I have witnessed too many unnecessary deaths here in the east. And, I fear, there will be more to come. It would appear, judging by what has preceded us, there will be significantly more dying than we can imagine. Come with me! We cannot outrun the tragedy that we have witnessed here. However, we can escape the immediate and continuous consequences likely to befall this army in the future."

"I trust you doctor and believe that your prophecy will, unfortunately, prove true. However, I can't leave this place just yet. The work that we have started has not been completed. Just recently, the following letter was brought to my attention.

Contraband Hospital, Washington, Nov. 5th, 1863

My dear Sister:

I shall depict our wants in true but ardent words, hoping to affect you to some action. Here are gathered the sick from the contraband camps in the northern part of Washington. If I were to describe this hospital it would not be believed. North of Washington, in an open, muddy mire, are gathered all the colored people who have been made free by the progress of our Army. Sickness is inevitable, and to meet it these rude hospitals, only rough wooden barracks, are in use-a place where there is so much to be done you need not remain idle. We average here one birth per day, and have no baby clothes except as we wrap them up in an old piece of muslin, that even being scarce. Now the Army is advancing it is not uncommon to see from 40 to 50 arrivals in one day. They go at first to the Camp but many of them being sick from exhaustion soon come to us. They have nothing that any one, in the North would call clothing. I always see them as soon as they arrive, as they come here to be vaccinated; about 25 a day are vaccinated. This hospital is the reservoir for all cripples, diseased, aged, wounded, infirm, from whatsoever cause; all accidents happening to colored people in all employs around Washington are brought here. It is not uncommon for a colored driver to be pounded nearly to death by some of the white soldiers. We had a dreadful case of Hernia brought in today. A woman was brought here with three children by her side; said she had been on the road for some time; a more forlorn, worn out looking creature

I never beheld. Her four eldest children are still in slavery, and her husband is dead. When I first saw her she laid on the floor, leaning against a bed, her children crying around her. One child died almost immediately, the other two are still sick. She seemed to need most, food and rest, and those two comforts we gave her, but clothes she still wants. I think the women are more trouble than the men. One of the white guards called to me today and asked me if I got any pay. I told him no. He said he was going to be paid soon and he would give me 5 dollars. I do not know what was running through his mind as he made no other remark. I ask for clothing for women and children, both boys and girls. Two little boys, one 3 years old, had his leg amputated above the knee the cause being his mother not being allowed to ride inside, became dizzy and had dropped him. The other had his leg broken from the same cause. This hospital consists of all the lame, halt, and blind escaped from slavery. We have a man & woman here without any feet, theirs being frozen so they had to be amputated. Almost all have scars of some description and many have very weak eyes. There were two very fine looking slaves arrived here from -Louisiana, one of them had his master's name branded on his forehead, and with him he brought all the instruments of torture that he wore at different times during 39 years of very hard slavery. I will try to send you a Photograph of him [where] be wore an iron collar with 3 prongs standing up so he could not lay down his head; then a contrivance to render one leg entirely stiff and a chain clanking behind him with a bar weighing 50 lbs. This he wore and worked all the time hard. At night they hung a little bell upon the prongs above his head so that if he hid in any bushes it would tinkle and tell his whereabouts. The baton that was used to whip them he also had. It is so constructed that a little child could whip them till the blood streamed down their backs. This system of proceeding has been stopped in New Orleans and may God grant that it may cease all over this boasted free land, but you may readily imagine what development such a system of treatment would bring them to. With this class of beings, those who wish to do good to the contrabands must labor. Their standard of morality is very low."

As John finishes reading he looks up and says, " These conditions persist in spite of all of our best efforts to end this war and the injustices which it has perpetuated and still will perpetuate, even after our time here. There is no evidence that the other side is willing to concede the righteousness of our cause. Nor, that those who have offered themselves as the supreme sacrifice, will not have died in vain. My people, and those that will follow them, still need me. My roots are still here and to leave now would be to abandon them. So I will have to bid you a good farewell and safe journey, at least for now and hope that our paths, as I believe they will, will cross once more in the future. If there is any justice in this world, we will not be deprived of each others friendship much beyond this time."

Post-Script

"Even if an army be defeated, it is better to have the supplies for proper care and comfort of the wounded upon the field, and run the risk of their capture, than that the wounded should suffer for want of them. Lost supplies can be replenished, but lives lost are gone forever."

Jonathan Letterman, Medical Recollections

Letterman's Improvements in Care and Treatment

The Letterman Plan is a system for treating and evacuating casualties from the battlefield developed during the American Civil War. Through his skill and insight Dr. Jonathan Letterman created the plan during the period when he served as Medical Director of the Union Army of the Potomac from July 1862 to the end of 1863. At this time he presided over U.S. medical operations at some of the most famous battles of the war, including Antietam, Fredericksburg, Chancellorsville, and Gettysburg.

When Letterman assumed command he found his department in great neglect and disorder. During his tenure, he not only strengthened medical operations but reconceived the task of caring for wounded soldiers. In so doing he created a comprehensive plan to handle mass casualties that synchronized the disparate elements of battlefield medicine in his day. It is a blueprint for medical operations that is still employed today.

The Letterman Plan is derived not from a single document, but from a series of reforms and programs instituted during his year and a half as Medical Director. For his efforts Jonathan Letterman is remembered as the father of modern battlefield and emergency medicine. The Letterman Plan today remains the foundation for elements of military medicine, as well as civilian emergency medicine and disaster relief.

When Letterman was appointed to the Army of the Potomac, it had suffered heavy casualties at Harrison's Landing on the James River where the army had retreated after the Peninsula campaign of 1862. Dr. Letterman's previous service on the frontier and in Indian expeditions had exposed him to the trials and tribulations of military life. It also gave him insight into the personal needs and requirements of the soldiers.

In response, Letterman introduced a variety of changes in supply and resource management. These changes followed those advocated by General McClellan. In a report on the subject McClellan wrote, "the nature of the military operations had also unavoidably placed the Medical Department in a very unsatisfactory condition. Supplies had been almost exhausted or necessarily abandoned; hospital tents abandoned or destroyed and the medical officers deficient in numbers or broken down by fatigue."

Despite many disadvantages and lax management at the time, Dr. Letterman focused most of his attention on the removal of the sick and wounded from the Peninsula. He was also instrumental in the creation of sanitary measures for improving and maintaining the health of the soldiers and on providing needed medical supplies in a timely and more effective manner. His orders were concise and practical and even his superior officer General McClellan praised Letterman's efforts in his report: "All the remarkable energy and ability of Surgeon Letterman were required to restore the efficiency of stations where patients were divided into categories according to the severity of their wounds and were treated accordingly. Letterman's system of organizing patients into those who would live regardless of their wounds and those who would die from their wounds is used to this day and has become known as triage in modern times.

Battlefields that were not yet introduced to the Letterman system were forced to tolerate inefficient and abusive systems of ambulance use. In one notorious example, on August 29, 1862 after the Battle of Second Bull Run it was discovered that three thousand wounded were left on the field for three days and six hundred were left for a week. This occurred because of negligence through widespread theft by the part of ambulance drivers who picked the pockets of the wounded, as well as stole alcohol from the medical supplies at their disposal and left the injured to die. As a response to these and other scandals, which he worked to eliminate, Letterman's system was gradually seen as an effective remedy to these faults and adopted over time by all of the Union's Armies.

Supply procurement and disbursal of the Army of the Potomac was also vastly improved. As the Army transferred from the Peninsula to Alexandria, Virginia, in 1862 Dr. Letterman discovered that supplies were fewer than required and deficient in quantity and availability. It was soon realized that the rapid and rushed transfer of the army caused supplies and ambulances to be lost or left behind.

The medical officers and officers of the Ambulance Corps were also very weary. Yet, the Army of the Potomac marched into Maryland and fought the battle of Antietam under these disadvantageous conditions in September of 1862. It was here that the value of Letterman's brilliant new Ambulance Corps system was demonstrated. After that battle, Dr. Letterman decided to make the method of getting medical supplies more efficient. He reduced the amounts of medicines and materials to be carried and reduced the number of wagons used to transport them by half of the previous numbers; thus he was able to convert the transport system into a more compact and functional apparatus of the army. The details of this new arrangement were published on October 4, 1862 and were republished on September 3, 1862. With the success of these improvements no further changes were ever found to be necessary. Adding to a new level of accountability Letterman instituted medical inspections and required detailed reports by his inspectors.

On October 30, 1862 Letterman established field hospitals while the Army of the Potomac was still in Maryland and this system was carefully designed to work with the Ambulance Corps and the method of supply as a whole. It was at the Battle of Fredericksburg that this holistic approach saw its first opportunity to prove its worth. Those who were a part of the conflict testified to the system's effectiveness. Surgeon Charles O'Leary, then Medical Director of the Sixth Corps said in his report: "Being appointed Medical Director of the Sixth Corps a few days prior to the Battle of Fredericksburg, December 13, 1862, I had the opportunity of putting in operation the Field-Hospital organization devised by the Medical Director of the Army of the Potomac, and witnessing its beneficial results. Within a very few hours after the positions were designated for the Field Hospitals on December 12, all the necessary appliances were on hand, and the arrangements necessary for the proper care of the wounded were as thorough and complete as I have ever seen in a civil hospital."

During the Battle of Gettysburg in July, Letterman had established a General Hospital and temporary field hospitals wherever there was a source of water and shelter. The buildings used ranged from churches, farm buildings and private homes. Sometimes the only shelter available to the operation was trees or a piece of canvas strung between poles. As the battle approached its end, there were approximately 23,000 wounded from both the Union and Confederate armies that required aid. It was a monumental task that the Medical corps faced, however, Letterman's system, properly implemented, saved the lives of many of these men.

According to the National Museum of Health and Medicine of the approximately 3 million soldiers and sailors who participated in Civil War combat approximately 750,000 died, where nearly 400,000 of these deaths were due to disease. (The figures in all cases are approximate.) The death toll was nearly 2% of the entire population. Among the 79% of combatants who survived the war, nearly half a million returned home permanently maimed or disabled. While most of these men sustained physical injury, a significant and unaccountable number also suffered what we now know as psychological trauma, later termed battle fatigue and even later, post-traumatic stress disorder.

Clara Barton

At the end of the war, Clara Barton continued her service to the soldiers and families directly affected by the war. In this task, she assisted grieving parents, family and friends whose many sons, brothers, neighbors had gone missing, presumably killed in the course of battle unacknowledged, and would be mourned in abstention.

In a bold gesture in 1865 Barton hired a staff and opened the "Office of Correspondence with the Friends of the Missing Men of the United States Army." As director she responded to over 63,000 letters, most of which required some kind of research that eventually lead to published lists of the names of the missing so that anyone with knowledge of their whereabouts or death could contact her. By the time the office closed in 1867, she had identified the fate of over 22,000 men.

Ms. Barton was an unusual person. In her varied service she gave to the men who inspired her keen devotion, disregarding the role of women, as defined at the time by men, as well as the restrictions placed upon her by army protocol and the traditions then in effect. She expressed her attitude toward the restraints that might detract her from her mission in the following statement:

"I have an almost complete disregard of precedent, and a faith in the possibility of something better. It irritates me to be told how things have always been done. I defy the tyranny of precedent."
Clara Barton

ABOUT THE AUTHOR

Stewart Cohen, PhD., taught for 36 years, in the Department of Human Development and Family Studies at the University of Rhode Island.

During his teaching career, he developed a special interest in the American Civil War, especially the men and women who lived during that turbulent era in our nation's history. Cohen's research and travels to American battle grounds helped him understand the crucial contributions of black people who had contributed to the war effort. Cohen found that often the involvement of black Americans in the Civil Was had only been footnoted by historians, many barely acknowledged, who had contributed to the war effort.

The stories in this book capture the contributions of those Black Americans as they lived, how their varied roles lead to greater freedom for all people, and the special facets and accomplishments that were characterized by the spirit of their lives. Stewart Cohen lives in Rhode Island with his wife Joan and her two cats, "Bonita" the Inquisitor and "Timi 2" the Bold.

OTHER BOOKS BY STEWART COHEN

ANTIETAM: AUTUMN BLIGHT: SLAUGHTER AT THE RIVER

FREDERICKSBURG: MASSACRE AT THE WALL

GETTYSBURG: DEATH AT THE ANGLE

SHILOH: BATTLE TO SAVE THE WEST

PETERSBURG: THE LAST SIEGE
(*COMING SOON*)

CHARLIE WRIGHT: THE SPY THAT SAVED GETTYSBURG
(*COMING SOON*)

www.ingramcontent.com/pod-product-compliance
Lightning Source LLC
Chambersburg PA
CBHW071311200626

46813CB00015B/1520